The Virtual Laboratory

Hans Meinhardt

The Algorithmic Beauty of Sea Shells

With contributions and images by Przemyslaw Prusinkiewicz
and Deborah R. Fowler

With 109 Illustrations, 96 in Color, and a $3^1/_2$" Diskette

 Springer

Hans Meinhardt
Max-Planck-Institut für Entwicklungsbiologie
Spemannstraße 35/IV
D-72076 Tübingen

Series Editor
Przemyslaw Prusinkiewicz

ISBN 3-540-57842-0 Springer-Verlag Berlin Heidelberg New York
ISBN 0-387-57842-0 Springer-Verlag New York Berlin Heidelberg

CIP data applied for

© Springer-Verlag Berlin Heidelberg 1995
Printed in Germany

Cover Design: Design Concept, Heidelberg
Typesetting by the author using LaTeX document preparation system
Data conversion: Lewis & Leins, Berlin
Printing and binding: Universitätsdruckerei H. Stürtz, Würzburg
SPIN 10115522   33/3142 – 5 4 3 2 1 0 – Printed on acid-free paper

It has turned out to be impossible … to get at the meaning of these marks … They refuse themselves to our understanding, and will, painfully enough, continue to do so. But when I say refuse, that is merely the negative of reveal – and that Nature painted these ciphers, to which we lack the key, merely for ornament on the shell of her creature, nobody can persuade me. Ornament and meaning always run alongside each other; the old writings too served for both ornament and communication. Nobody can tell me that there is nothing to communicate here. That it is an inaccessible communication, to plunge into this contradiction, is also a pleasure.

Thomas Mann, Doktor Faustus, III. Chapter: Jonathan Leverkühn contemplating a pattern on a New Caledonien sea shell. After the translation from the German by H. T. Lowe-Porter, Penguin Books.

The same text in the original version of Thomas Mann:

Es hat sich … die Unmöglichkeit erwiesen, dem Sinn dieser Zeichen auf den Grund zu kommen. … Sie entziehen sich unserem Verständnis, und es wird schmerzlicher Weise wohl dabei bleiben. Wenn ich aber sage, sie entziehen sich, so ist das eben nur das Gegenteil von "sich erschließen", und daß die Natur diese Chiffren, zu denen uns der Schlüssel fehlt, der bloßen Zier wegen auf die Schale ihres Geschöpfes gemalt haben sollte, redet mir niemand ein. Zier und Bedeutung liefen stets nebeneinander her, auch die alten Schriften dienten dem Schmuck und zugleich der Mitteilung. Sage mir keiner, hier werde nicht etwas mitgeteilt! Daß es eine unzugängliche Mitteilung ist, in diesen Widerspruch sich zu versenken ist auch ein Genuß.

Preface

The pigment patterns on tropical shells are of great beauty and diversity. They fascinate by their mixture of regularity and irregularity. A particular pattern seems to follow particular rules but these rules allow variations. No two shells are identical. The motionless patterns appear to be static, and, indeed, they consist of calcified material. However, as will be shown in this book, the underlying mechanism that generates this beauty is eminently dynamic. It has much in common with other dynamic systems that generate patterns, such as a wind-sand system that forms large dunes, or rain and erosion that form complex ramified river systems. On other shells the underlying mechanism has much in common with waves such as those commonly observed in the spread of an epidemic.

A mollusc can enlarge its shell only at the shell margin. In most cases, only at this margin are new elements of the pigmentation pattern added. Therefore, the shell pattern preserves a record in time of a process that took place in a narrow zone at the growing edge. A certain point on the shell represents a certain moment in its history. Like a time machine one can go into the past or the future just by turning the shell back and forth. Having this complete historical record opens the possibility of decoding the generic principles behind this beauty.

My interest in these patterns began with a dinner in an Italian restaurant. During the meal I found a shell with a pattern consisting of red lines arranged like nested W's. Since I had worked for a long time on the problem of biological pattern formation, this pattern caught my interest, more from curiosity. To my surprise it seemed that the mathematical models we had developed to describe elementary steps in the development of higher organisms were also able to account for the red lines on my shell. Thus, the shell patterning appeared to be just another realization of a general pattern forming principle. But this observation did not remain unchallenged for long. Soon thereafter I saw the complexity and beauty of tropical shells and realized that these patterns are not explicable on the basis of the elementary mechanisms in a straight forward manner.

We do not know what these patterns are good for. Presumably there is no strong selective pressure on the shell pattern. Variations are possible without severely influencing the viability of the animals. Since, as will be described in this book, the patterns result from the superposition of several pattern-forming reactions, their diversity provides a natural picture book to study complex non-linear pattern formation.

To find models for these complex patterns turned out to be much more difficult than I thought. Of course, before making a simulation I was convinced that I had found the correct model. Using the simulation I learned frequently where mistakes in my thinking were and to what patterns my hypothesis really would lead. This led to new insights and new models. I am far from having a satisfactory model for every shell. But I hope that this book invites you to search for alternative and new solutions.

The book is accompanied by a computer program for performing the simulations on a Personal Computer. Most simulations shown can be reproduced. To see the emergence of these patterns on the screen provides a much more intuitive feel for the dynamics of the system. Since minor fluctuation can play a decisive role, even the repetition of the same simulation can lead to a somewhat different pattern. This corresponds to the fact that the patterns on any two shells are never identical. The program allows you to change parameters such as the life time of a substance or its spread by diffusion. The consequences of these changes can be seen immediately as an alteration of the pattern. The program is provided with full source code (Microsoft Professional, Quick or Visual Basic, Power Basic). Therefore, new model interactions can be easily inserted.

Acknowledgements

This book would not have been completed without encouragement from many quarters, foremost from my wife Edeltraud Putz-Meinhardt. I would like to express my thanks to those who contributed to the book. The basic ideas grew out of a theory I developed with Alfred Gierer. His concept of local autocatalysis and long ranging inhibition has been the basis of most of my work on biological pattern formation. Martin Klingler described in his master's thesis many interactions capable of reproducing shell patterns in fine detail. Discussions with Andre Koch and Kai Kumpf have been stimulating for me. I thank Drs. Ellen Baake, Jon Campbell, Christa McReynolds, Arthur Roll, Adolf Seilacher and Ruthild Winkler-Oswatitisch for shells and photographs. Christa Hug helped to prepare the manuscript, Karl Heinz Nill made the drawings, and Dr. Hans Wolfgang Bellwinkel brought the Thomas Mann-quotation to my attention. I am very grateful to Deborah Fowler and Przemyslaw Prusinkiewicz from the University of Calgary, Canada, for their contribution of the chapter on shell shapes (chapter 10) and Lynn Mercer for her very careful correction of the manuscript.

Last but not least, I am most grateful for the excellent working conditions provided by the Max-Planck-Institut für Entwicklungsbiologie in Tübingen over many years.

Hans Meinhardt

Contents

Chapter 1

Shell patterns – a natural picture book to study dynamic systems and biological pattern formation

1.1 Dynamic systems everywhere

Everyday we are confronted with systems that have an inherent tendency to change. The weather, the stock market, or the economic situation are examples. Dramatic changes can be initiated by relatively small perturbations. In the stock market, for instance, even a rumour may be sufficient to trigger sales, lowering quotations and causing panic reactions in other shareholders.

An essential element of dynamic systems is a positive feedback that self-enhances the initial deviation from the mean. The avalanche is proverbial. Cities grow since they attract more people, bacteria or viruses can replicate and the progeny start replicating too. In the universe a local accumulation of matter may attract more dust, eventually leading to the birth of a star.

Earlier or later self-enhancing processes evoke antagonistic reactions. A proliferating virus may trigger an immune response that neutralizes the virus. A collapsing stock market stimulates the purchase of shares at a low price, thereby stabilizing the market. The increasing noise, dirt, crime and traffic jams may discourage people from moving into a big city.

In addition to the balance between self-reinforcing and antagonistic tendencies, several other elements play a decisive role in the fate of dynamic systems. For instance, if the antagonistic reaction follows with some delay, the self-enhancing reaction can cause an overshoot or even an explosion. The explosion of dynamite is a good example. After ignition of a small portion the resulting heat and the shock wave ignite more of the explosives in the neighbourhood. The reaction is so vehement because the oxygen required for burning is part of the chemical and is available immediately. Thus, no antagonistic effect slows down the reaction until the explosive is used up. Of course, afterwards a further ignition is impossible.

But let us consider a fire. Fire is also a self-enhancing process since more heat releases more burnable gases from the fuel. But the depletion of oxygen may represent an antagonistic reaction that keeps the fire down to the point that only smouldering is possible. In such a case, the rapid antagonistic reaction, the oxygen depletion, hinders the development of a big fire. The burning process can go on for a much longer period although at a lower level. Thus the ratio of reaction times between the self-enhancing and the antagonistic processes plays a decisive role.

Another example should illustrate the same fact. As a rule, it takes about two days to fully develop an influenza but it takes about a week to get rid of it. Thus, it appears that our immune system responds too slowly when compared with the growth rate of the virus. Initially the virus proliferates in an avalanche-like manner and we become sick. But what appears at first as a misconstruction turns out to be an advantageous strategy. The slower responding immune system accumulates more and more specific antibodies until the entire virus can be trapped. The body can completely rid itself of the virus. If the immune system responded much faster, a balance between the proliferating virus and the immune system would be established at a lower level. The body would have to fight for the rest of its life against the ever proliferating virus since partial removal of the virus would lead to a corresponding down-regulation of the immune response, providing a new chance for the virus. With the system as it is, we are sick for a week, but after this week we are healthy again and free of the virus.

1.2 Pattern formation

Another decisive parameter in a dynamic system is the spread of its components. In the example mentioned above, the virus may be transmitted to another person who will also become sick after some delay. The infection spreads like a travelling wave. This spread is possible only since the self-enhancing agent, the virus, but not the antagonistic reaction, the immune response, can be transmitted to another individual.

In other systems, it is the antagonistic reaction that spreads more rapidly, and this can lead to stable patterns. Let us regard the formation of sand dunes in the desert (Figure 1.1). Dunes are formed despite the fact that the wind very quickly redistributes the sand. Dune formation may be initiated by a stone in the desert that provides a wind shelter. Sand accumulates behind the wind shelter, and a dune begins to grow. But the sand, once settled in the dune, cannot participate in dune formation somewhere else. The growth of a dune lowers the sand content in the air. The antagonistic reaction results from this removal of sand particles being moved by the wind, and has a long range effect. In this way, the probability of initiating new dunes and the growth of existing dunes in the surrounding area is reduced. In contrast, the increased accumulation of sand behind the wind shelter has a range comparable with the size of the dune. Thus, the basic elements for the formation of stable patterns are a short range self-enhancing reaction and a long range antagonistic reaction.

A similar argument can be made for the formation of valleys and rivers by erosion. The pattern of a ramified river system is certainly not preceded by a corresponding pattern of rain fall, but results from a self-organizing process. The

Figure 1.1. The sand dune paradox. Naively, one would expect that the wind in the desert causes a structureless distribution of the sand. However, wind, sand and surface structure together represent an unstable system. Sand deposits more rapidly behind a wind shelter. This increases the wind shelter which, in turn, accelerates the deposition of more sand – a self-enhancing process.

deepening of a valley proceeds essentially by the erosion of a meandering river. A larger valley collects water from a more extended portion of the surface. Thus, a larger valley has a better chance of becoming even deeper.

The antagonistic reaction can result from a depletion of material trapped by the self-enhancing process, such as in the sand dune example mentioned above. Alternatively, a direct inhibitory effect may spread out from such a self-enhancing center. In the formation of stars both effects play a role. A local increase of matter attracts more cosmic material – the self-enhancing process. One of the antagonistic reactions results from the depletion of cosmic dust in the surroundings. In addition, there is an active antagonistic effect produced by the developing star: the emitted light exerts a so-called light pressure that repels dust particles.

1.3   Dynamic systems are difficult to predict

Investigations of so-called chaotic systems have emphasized the fact that processes exist that are inherently unpredictable on the long term although each step in the development is unequivocally determined by the preceding situation. The weather is an example. Calculation of future developments would require the knowledge of a given situation with an arbitrary precision – a knowledge that is impossible to obtain.

The situation is similar in the systems discussed above where strong positive feedback couplings are involved. Minute differences in the initial conditions can cause a completely different outcome if the situation is just on the border at which a self-enhancing process becomes dominant. If self-enhancement is triggered in such a "revolutionary" situation, it will obtain a dynamic fairly independent of the mode of ignition.

If small differences are responsible for the selection of different pathways, our intuition of such systems is unreliable. Many attempts have been made to obtain a better understanding using mathematical modelling. For complex systems this is possible only using approximations since one can never be sure that all critical parameters have been considered. The problem becomes even more severe if the feelings and thinking of human beings have a strong influence on the fate of a system, as it is the case in politics or economy. The difficulty in making a prediction is obvious if even a rumour can induce a panic that is not justified by the real situation.

There are two main reasons for modelling dynamic systems: to provide a check on whether a system is fully understood, and to make predictions, at least for the near future.

To obtain the laws that govern a system, a comparison with its development in the past is an important check on whether system is correctly understood, at least in retrospect. This requires good historical data.

Due to the importance of dynamic systems on the one hand and the difficulty of understanding them on the other, it seems advisable to study relatively simple model systems. A very particular model will be discussed in the present book – the patterns on the shells of molluscs. These motionless calcified patterns are more reminiscent of artistic decorations on china than dynamic systems. However, a closer inspection of many shells reveals dramatic events in their history.

1.4   Pattern formation in biology

The generation of patterns on the shells of molluscs is, of course, only a very special case of the general problem of how an organism obtains its complex structure during development. The life cycle of a higher organism starts, as a rule, with a single fertilized cell. At the end of embryonic development a very sophisticated arrangement of highly specialized cells is generated. The similarity of identical twins is an indication of how stringent this process is under genetic control. But reference to genes does not provide an explanation of this process *per se* since, as a rule, with each cell division both daughter cells obtain the same genetic information.

It appears to be a hopeless enterprise to find a mathematical description (and thus an unequivocal understanding) of a process as complex as the formation of a higher organism. However, it turns out that this process can be separated into many steps that can be regarded, as a first approximation, to be independent of each other. For instance, a very important process for a developing embryo is the formation of the primary embryonic axes. Some signals must be present to determine where to form a head, tail and so on in the initially more or less uniform (usually hollow) sphere of cells. It turns out that for the fruit fly *Drosophila*, for instance, the anteroposterior axis (head to tail) is under the control of a completely different set of genes than the dorsoventral (back to belly) axis (Nüsslein-Volhard 1991). Another example of the independence of a structure from the surrounding tissue is the formation of legs. After its initiation, an amphibian leg develops fairly normally even after transplantation to an ectopic position. Its development proceeds under the control of a local coordinate system. Because of this partial independence one can make models of the individual elementary steps. Of course, the steps must be linked together in order to position the individual structures in correct relation to each other. For instance, in a developing organism it is essential that the head-to-tail and the back-to-belly axes are arranged perpendicular to one another, but this is already a refinement.

We have proposed several models of biological pattern formation for specific developmental situations (Gierer and Meinhardt, 1972, Gierer, 1981, Meinhardt, 1982). It came as a surprise to me that the patterns on the shells of molluscs could be described with basically the same equations that were initially derived to describe elementary steps in biological pattern formation, such as the formation of embryonic axes, the head formation of the freshwater polyp *Hydra*, or the initiation of periodic structures such as leaf formation at the tip of a growing shoot. Thus, the shell patterns will be used as a natural picture book to become more familiar with a general mechanism that is the basis for a very important process, biological pattern formation.

1.5 Most shell patterns preserve a faithful time record

In normal development, a strong evolutionary pressure exists to reproduce a given structure faithfully. Moreover, a structure, once formed, usually remains stable at least for a certain time interval. In contrast, the functional significance of the pigment patterns on shells is not clear. Many molluscs live buried in the ground, some are covered with an opaque layer, the periostracum. Thus, presumably there is no strong selective pressure on the shell patterns. The diversity indicates that it is possible to modify the pattern drastically without endangering a species. Nature allowed to play.

Shells consist of calcified material. The animals can increase the size of their shells only by accretion of new material at the margin, the growing edge. In Figure 1.12 later in this chapter the edge of a shell is clearly to be seen. Most decorations of shells result from the incorporation of pigments during this growth process. Once made, as the rule, the pattern remain unchanged. The patterns are therefore historical records of what happens at the growing edge, i.e., they are a time record of a pattern forming process in a more or less linearly arranged array of cells.

Figure 1.2. Different modes of pattern formation on shells of *Cypraea diluculum*. The banding pattern results during the growth of the shell by the sequential addition of new material. The pattern is a time record of a linear pattern forming process taking place at the growing edge. The dot-pattern around the opening results from a two-dimensional pattern forming process at later stages. The snail engulfs its shell by an ectodermal protrusion in which the corresponding pattern is generated. Pigment produced in this layer becomes deposited on the shell.

Some shells, however, produce their pattern in a totally different way. Ectodermal protrusions engulf the shells and pattern forming processes within these sheets are copied onto the shell. Thus, in these cases pattern formation results from a two-dimensional process. The pattern represents a snapshot of a particular moment, not a time record. Some snails change from one mode to the other after reaching adulthood. On the shell of *Cypraea diluculum* (Figure 1.2) one can see both types of patterns on the same shell. While the juvenile pattern consist of oblique lines, the later pattern formed around the shell opening consists of isolated dots. Although, the patterns look so different, we will see that they can be explained by basically similar mechanisms.

For the purpose of the book, the first type of patterns, those that are generated over the course of time, are most interesting since they bear the historical record of their formation. They provide therein a key for deciphering the underlying pattern forming process. I will deal mostly with this type of pattern.

1.6 Elementary patterns: Lines perpendicular, parallel and oblique to the direction of growth

Shells show an enormous diversity of patterns whereby related species can show very different patterns while non-related species can show very similar patterns. The shells shown in Figure 1.3 provide an example: one shell belongs to a snail, the other to a bivalved mussel. Both shells show oblique lines that branch. They share a

a
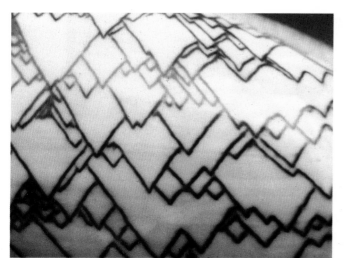
b
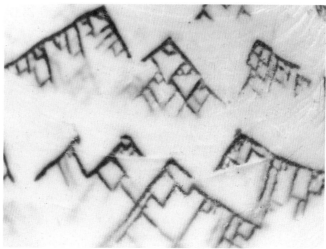

Figure 1.3. Similar pattern on non-related molluscs. (a) Detail of the shell pattern of *Oliva porphyria*, (b) the bivalved mussel *Lioconcha hieroglyphica*. Both patterns consist of branching oblique lines. The branching occurs frequently at the same level on distant lines indicating that branch initiation is a non-local process (see Figure 6.1).

very characteristic element. The initiation of several branches occurs frequently at the same horizontal level, i.e., it can take place simultaneously at distant positions, a feature that will be discussed later in detail. The similarity in non-related species indicates that the different patterns are generated by a common mechanism and that the diversity is generated by minor modifications.

To find an inroad into the logic behind these patterns, it is advisable to start with more elementary patterns. Figure 1.4 shows shells with lines parallel and perpendicular to the direction of growth. Keeping in mind the space-time character of the shell pattern, lines parallel to the direction of growth indicate the formation of a spatial periodic pattern of pigment production along the edge that is stable in time. At more or less regular distances, groups of cells in the mantle gland produce permanent pigment while cells in between never do so. This is the usual situation in morphogenesis where particular structures such as leaves, hairs or feathers become initiated at regularly spaced positions.

Other patterns indicate that pigment deposition oscillates. A particular cell produces pigment only during a certain time interval and then enters into an inactive (refractory) period until the next pigment producing phase occurs. A synchronous oscillation in pigment production leads to stripes parallel to the axis. In the example given in Figure 1.4 it can be clearly seen that the oscillations are almost but not completely synchronous. Some regions of the edge are white, others are pigmented. Thus, the synchronization cannot be a result of external influences, like daily or seasonal fluctuations, in contrast to the situation, for instance, in tree rings.

Figure 1.4. Elementary pattern: stripes perpendicular and parallel to the direction of growth. In the upper shell pigmentation has occurred at regular time intervals more ore less synchronous. Axial stripes (i.e., stripes perpendicular to the direction of growth) results. In the lower shell pigmentation occurred permanently at regularly spaced positions leading to stripes parallel to the direction of growth.

The mechanisms that lead either to synchronous oscillations or to a stable pattern in space cannot be very different from each other. The specimen at the bottom of Figure 1.4 has on its cone a pattern resulting from oscillations while the main pattern was generated by a stable system. As shown below, a change of the ionic strength of the water is already suffcient to cause a corresponding pattern alternation (see Figure 1.11).

1.7   Oblique lines

Oblique lines originate from travelling waves of pigment production. Such waves arise if pigment-producing cells trigger their neighbouring cells so that – after a certain delay – these cells also start to produce pigment and so on, analogous to the formation of the influenza wave mentioned earlier.

Pure elementary patterns – stripes parallel, perpendicular or oblique to the growing edge – are more the exception than the rule. For instance, the perpendicular lines in the lower shell of Figure 1.4 show small gaps at regular intervals

Figure 1.5. Bulges instead of pigmentation: elementary patterns of the relief type. Ripples parallel, perpendicular and oblique to the direction of growth. They increase the friction of the shell in the sand or mud.

suggesting that an oscillating pattern was superimposed. The branching of the oblique lines shown in Figure 1.3 indicates that the travelling waves involved in shell patterning can have unusual properties. At a certain moment, a wave can split producing a wave that moves backwards – a process that never occurs, for instance, in a nerve pulse that travels along a nerve fibre.

1.8 Relief-like patterns follow the same rules

Pigmentation is not the only possible decoration on shells. Relief-like structures are also very frequent. The same elementary patterns as listed above occur. Figure 1.5 shows corresponding examples.

As a rule, relief-like patterns are less complex and more reproducible from specimen to specimen. Relief-like patterns also have biological functions. For bivalved mussels, the rough surface increases the friction with the sand during the opening and closing of the two shells and facilitates in this way burrowing into the sand or mud. As shown by paleontologic work (Seilacher, 1972), species that originally appear different can develop towards similar shapes and surface structures if they populate the same habitat and are forced to behave in a similar way. Therefore, in contrast to pigmentation patterns, relief-like patterns seem to be shaped by strong selective pressure.

1.9 Many open questions and some hints

Very little is known about the mechanisms that lead to shell patterning. The models worked out in this book can only describe which type of interaction could in principle account for such patterns. No information can be obtained about the chemical nature of the substances involved. Sometimes two different models can reproduce the natural counterpart with reasonable agreement. Usually, however, related species with similar patterns are available that can be described much better by one or the other mechanism. Therefore, the model's handling of natural variability provides one of the criteria for its quality.

Some shells show very particular features that are helpful in getting ideas about the underlying mechanism. Some examples will follow.

Occasionally shells show an obvious perturbation of the normal pattern. A traumatic event must have happened in the animal's history such as temporary dryness, lack of food or injury by a predator. After such a perturbation, the pattern may be very different for a long period. There is strong support for a model that is able to account not only for the normal pattern but also for the pattern regulation after such a perturbation. Several examples will be given later on.

Since the shell grows in several rounds around the axis, parts of the progressing shell formation are in direct contact with parts formed in a previous round. On some shells oblique lines extend from an older (inner) region to a newer (outer) winding without major discontinuity (Figure 1.6). This indicates that an existing pigmented region, if touched by a growing mantle gland, can initiate a travelling wave of pigment production on the newly formed portion of the shell. The result is that both stripes appear in register. These snails can "taste" the old stripe. Such tasting has been proposed by Ermentrout et al. (1986) for other reasons, namely to account for the sometimes very long periods in the oscillation of pigment depositions.

A very interesting phenomenon can be seen on the right shell of Figure 1.6. There, only every second oblique line is in register with an old line. Obviously, the spontaneous oscillation frequency was too high during the last round of growth. Only every second initiation of pigment production could be triggered by the old pattern while in between a spontaneous trigger took place.

Similarly, on bivalved molluscs, a synchronization can take place between shells such that the two shells become mirror-symmetric to each other. Figure 1.7 shows two pairs. Obviously, some cross-talk took place between the two shell-producing mantle glands. If one produces pigment, the other produces pigment too. This, however, takes place only if relatively coarse patterns are generated. If more detailed patterns are formed, both patterns may have common features but the details may be different. Figures 5.1 and 5.8 show two complementary and more complex shells.

Sometimes it is difficult to decide what is pattern and what is background. The two shells in Figure 1.8 show dots. One shell carries pigmented dots on

Figure 1.6. An existing pigment pattern can trigger the formation of a new one. Oblique lines are frequently in register on parts of the shell that have been formed during different rounds of shell growth. The arrows mark some instances. This can only be interpreted by the assumption that the mantle gland can detect existing pigmentation and that this can initiate a travelling wave. In the right shell only every second line continues (arrows).

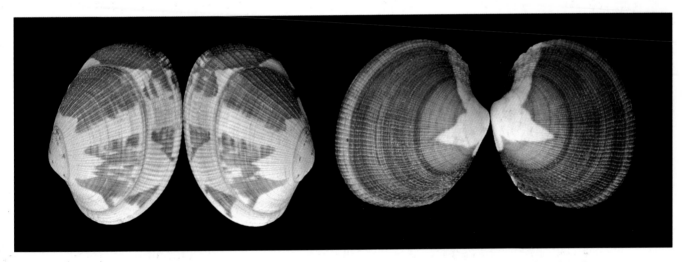

Figure 1.7. Mirror-symmetry in bivalved shells of molluscs. Some cross-talk must take place between the two shell-producing glands such that pigment formation on one shell triggers pigment formation on the other. Finer patterns have common features but are different in the details.

Figure 1.8. Figure and ground: Staggered dots in reverse pigmentation on *Babylonia papillaris* and *Heridina natalensis*. Finer details, for instance the dark border at the lower edge of the white patches, indicate that the two patterns depend on different mechanisms.

less-pigmented background; on the other shell the situation is the reverse. The similarity could indicate that a common mechanism generates a signal, but the two molluscs make different use of that signal. In one case the signal promotes, and in the other case the signal inhibits pigmentation.

However, a detailed inspection reveals more differences. The unpigmented patches on the left shell show narrow, very dark borders at their lower edge and the pigmented background has a fine structure of densely packed narrow lines. It is an extreme form of a general pattern that will be treated in detail further on (see Figure 9.7). The common feature is the offset of a particular pattern element along the space and time coordinates. But this feature can be achieved by several mechanisms that differ in their degree of complexity.

Other shell patterns indicate that two interacting systems are involved. For instance, on the shell of *Conus puncticulis zeylanicus* (Figure 1.9) rows of dark, crescent-like patches are visible. The patches within a row are at very irregular distances from each other. A closer inspection reveals that the irregularity results from a random alternation of the pigmented patches with patches that are significantly less pigmented than the light gray background. Thus, two signals seem to be produced, one that causes pigmentation and another one that suppresses pigmentation. The irregularity excludes the idea that the dark and white patches result from a single oscillating system at different stages of the cycle. However, some coupling between the two systems must exist since the dark and the white spots always keep a certain distance from each other.

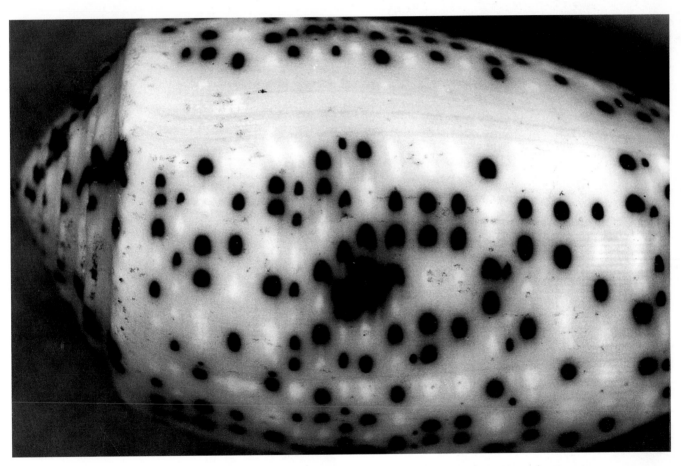

Figure 1.9. Irregular alternation of dark pigmentation and pigmentation lighter than the gray background: *Conus puncticulis zeylanicus*. This pattern indicates the involvement of two different signal systems, one that enhances and another that suppresses pigmentation.

Figure 1.10. Global oscillations and pigmentation pattern. The modulation of the background in the left shell indicates a global oscillation that controls or synchronizes the dark pigmentation, causing the ladder pattern. In the right shell some pigmentation lines are occasionally shifted against each other. This indicates the absence of global control. Also no background modulation is visible. Thus, even in related species the pattern forming mechanism can be different. The synchronizing influence of a global oscillation on pigment deposition is responsible for several other phenomena (see Figure 4.14).

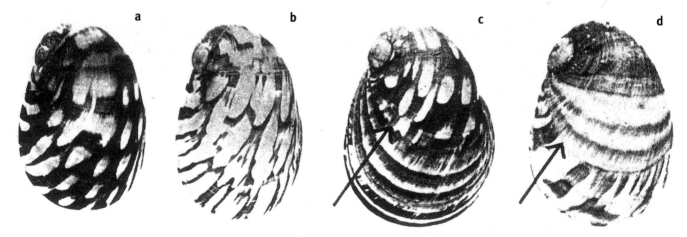

a   b   c   d

Figure 1.11. Modification of shell patterns by ionic strengths. (a, b) Variants of a fresh water snail with dense and sparse shell pigmentation, kept at low ionic strengths. Shift of an animal in the aquarium (c) from freshwater to high (0.37%) or (d) from high to low salt concentrations causes a dramatic pattern change. Arrows indicate the time of change. (Photographs kindly supplied by D. Neumann, see Neumann, 1959b)

Another example where a faint pigmentation provides some hints of the mechanism is given in Figure 1.10. Parallel lines are framed by oblique lines, providing an overall impression of connected triangles. A closer inspection of the yellowish background reveals that there is a modulation in the same phase as the dark horizontal lines. Obviously, a global oscillation exists that acts either as a precondition or it modulates the oscillation that causes the darker pigmentation. As a note of precaution, the second shell of Figure 1.10 is a related species with a similar pattern. In this shell, however, the horizontal lines are partially out of phase. No synchronization by a global oscillation and no modulated background can be seen. Therefore, even in related species it is dangerous to generalize too much.

Little experimentation has been reported with shell patterns. The pattern of the freshwater snail *Theodoxus fluviatilis* depends on the salt concentration (Neumann, 1959a-c). In Germany, waste water with high salt concentrations from potassium mining is introduced into the river Werra. He found pattern changes downstream from the waste water dispersal. These pattern changes can be reproduced in the laboratory. Figure 1.11 shows the pattern after a shift from high to low salt concentrations and *vice versa*. Obviously, there is no strong regulatory system in the animal to maintain its typical pattern. To the contrary, all the examples mentioned above indicate that the pigment producing system is very sensitive. A change in the external conditions, an influence from pigmentation laid down earlier or from the shell counterpart can modify pigment production.

1.10 The hard problem: complex patterns

Many shells show highly complex patterns. The shell of *Conus textile*, Figure 1.12, provides an example. The complexity is not without rules. Some regions show a light brown, others a white background. Darker oblique lines are visible whose general character depend on the background, more faint in the white but thicker in the light brown region. Moreover, there is a strong tendency to change from parallel lines (synchronous oscillation) in the brown region to oblique lines (travelling waves) that branch in the white region. Keeping the space-time character in mind, it is clear that the transition from white to brown is preceded by a dark brown line, while the brown-white transition is not. The faint oblique lines in the white region frequently have their origin in a thick dark line in the brown region. The brown-white transition occurs usually simultaneously in an extended region, causing a light brown-white border parallel to the edge.

This complexity cannot result from a single pattern-forming reaction. Two or more reactions must be superimposed that influence each other. Usually only one pattern, the pigmentation, is visible. A second pattern that modifies the pigmentation pattern may be invisible but must be deduced from the unusual behaviour of the pigmentation pattern. In the case of *Conus textile* the light brown pattern is presumably an exception in that a visible trace of the second pattern system

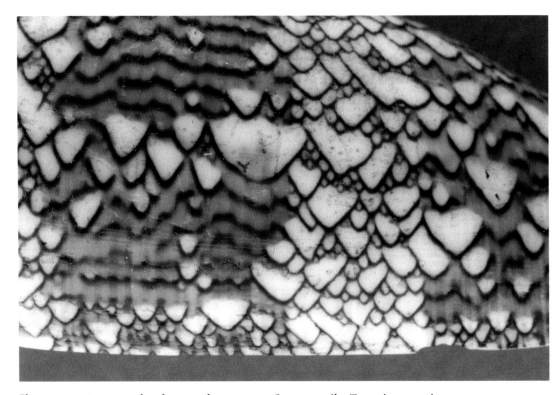

Figure 1.12. An example of a complex pattern: *Conus textile*. Two pigmentation systems appear to be superimposed. Light brown pigmentation occurs preferentially in two bands that cover large regions. In addition oblique lines with a much darker pigmentation exist. The dark pigmentation system is influenced by the light one. In the light brown region the dark brown pigmentation lines are much thicker and are preferentially oriented parallel to the growing edge. In regions with non-pigmented background the dark lines are much narrower.

exists. The different behaviour of the pigment system – synchronous oscillations *versus* travelling waves depending on whether the light brown pattern is in the ON or OFF state – suggests general modifications of one system by the other. The problem in understanding complex patterns lies in the enormous number of combinatorial possibilities between two or more systems. Each component of one system can activate or suppress another; the influence may involve changes in the production or destruction rates; and so on. The simulation of the complex patterns provided in this book should be regarded only as an attempt to decipher the complex interaction and as an invitation to search for other possibilities.

1.11   Earlier attempts to understand shell patterns

Several attempts to model patterns on mollusc shells have been made. Formal models have been proposed by Waddington and Cowe (1969) for the tent-like patterns of Oliva shells and by Lindsday (1982) for bivalved molluscs. Cellular automata models have been discussed by Herman and Liu (1973), by Wolfram (1984) and by Plath and Schwietering (1992). Wanscher (1971) tried to explain shell patterning by a shielding mechanism between the pigment-producing mantle gland and the periostracum, the side of pigment deposition. The white drop-like non-pigmented regions in Conus textile (Figure 1.12) or in Conus marmoreus (Figure 7.2) have been discussed as an example. The idea of using reaction-diffusion mechanisms to model shell patterns dates back to 1984 (Meinhardt, 1984). A model based on similar interactions but emphasizing the role of the nervous system was proposed by Ermentrout, Campbell and Oster (1986; see also Murray, 1989). Many of the ideas outlined in this book were first published by Meinhardt and Klingler (1987, 1991).

In the next two chapters, a general theory of biological pattern formation will be outlined and applied to shell patterning. The elementary patterns result from a straightforward application of this basic theory. In subsequent chapters extensions of the elementary mechanisms will be discussed that account for the many unusual features of these beautiful pattern forming systems.

Figure 2.1. Patterns stable in time and their traces on shells. Stripes perpendicular to the growing edge, i.e., parallel to the direction of growth, result from stable pigment production at a particular position and its suppression at locations in between. In the lower shell, different sets of stripes are formed on the inner and the outer surface of the shell. On the outer pattern, the stripes are formed in pairs. An explanation for this phenomenon will be given in Figure 2.4f.

Chapter 2

Pattern formation by local self-enhancement and long range inhibition

Like other biological processes pattern formation is based on the interaction of molecules. In order to find a mathematical description for a particular process the concentrations of the substances involved must be described as a function of space and time. This is possible by using equations that describe the *changes* in concentration over a short time interval as a function of other substances. Adding these concentration changes to given initial concentrations provides us with the concentration at a somewhat later time. Repetition of such a calculation provides the total over the course of time. Three factors are expected to play a major role in the concentration change: the rate of production, the rate of removal (or decay), and the loss or gain due to an exchange with neighbouring cells, for instance by diffusion.

As mentioned earlier, we have proposed that pattern formation starting from initially more or less homogeneous conditions requires local self-enhancement coupled with a long range antagonistic effect (Gierer and Meinhardt, 1972, Gierer, 1981, Meinhardt, 1982). Patterns are formed because small deviations from a homogeneous distribution create a strong positive feedback which causes the deviations to grow even more. A long-range antagonistic effect restricts the self-enhancing reaction and causes a localization.

2.1 The activator – inhibitor scheme

The scheme of a biochemically feasible realization of this general principle is shown in Figure 2.2. A short-range substance, the activator, promotes its own production (autocatalysis) as well as that of its rapidly diffusing antagonist, the inhibitor. The concentrations of both substances can be in a steady state. A general increase in the activator is compensated by an increase in the inhibitor concentration. However, such an equilibrium is locally unstable. Any *local* increase of the activator will increase further due to autocatalysis despite the fact that a surplus of inhibitor is also produced by this local increase. It diffuses rapidly into the surroundings and slows down the autocatalysis while the local activator elevation increases further (Figure 2.2). A set of partial differential equations is given in Equation 2.1 (see

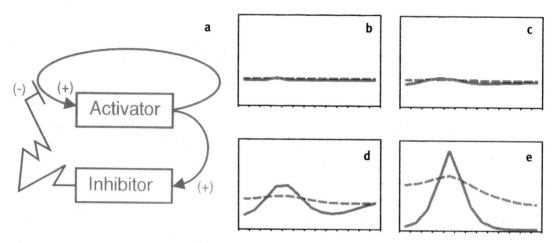

Figure 2.2. Pattern formation by autocatalysis and long-range inhibition. (a) Reaction scheme. An activator catalyses its own production and that of its highly diffusing antagonist, the inhibitor. (b-e) Stages in pattern formation after a local perturbation. Computer simulation in a linear array of cells. A homogeneous distribution of both substances is unstable. A minute local increase of the activator (—) grows further until a steady state is reached in which self-activation and the surrounding cloud of inhibitor (- - - -) are balanced [S22].

box on page 23)[1]. In terms of these equations the crucial condition for pattern formation is that the diffusion of the inhibitor is much higher than that of the activator, i.e., the condition $D_b \gg D_a$ must be fulfilled. As shown by Granero *et al.* (1977) the inhibitor must diffuse at least seven times faster. A simple calculation is given in Equations 2.2a-c (page 24) that should provide some intuition into why this interaction can generate stable patterns starting from nearly homogeneous initial conditions.

2.2   Stable patterns require a rapid antagonistic reaction

Patterns that are stable in time for a certain period are most common in biological pattern formation. They are required, for instance, in the early development of an embryo when the anteroposterior (head to tail) and the dorsoventral (back to belly) axes are determined. These signals may disappear once their purpose is fulfilled, but they should not reappear in a periodic manner in different positions. On shells, stable patterns lead to permanent pigment production in some positions and its suppression in between. This leads to an elementary pattern of stripes parallel to the direction of growth. Examples are given in Figure 2.1. In normal biological pattern formation a strong selective pressure exists to maintain a decision once it is made, for instance to form a head at a particular position. No such preference

---

[1]The names of parameters are partially changed from our previous publications in order to use identical names in the equations and the computer program (see chapter 11).

for stable patterns exists on shells. Therefore these stripes are only one pattern among many others.

In order for a pattern to become stable over time the antagonist must react very quickly to changes in activator concentration. Otherwise oscillations will occur, a mode that is also of special significance for shell patterning (see chapter 3). A rapid adaptation of the antagonist regulates any deviation from a steady state concentration. In the activator-inhibitor scheme this means that the inhibitor must have a higher turn-over rate than the activator in order for a rapid adaptation of the inhibitor concentration to be possible. In terms of Equation 2.1, the rate of decay or removal of the inhibitor r_b must be larger than that of the activator, *i.e.* the condition $r_b > r_a$ must be satisfied for stable patterns.

In the following section, the properties of reactions under these conditions will be discussed. At the end of the chapter, other molecular feasible realizations using the same principle will be given.

The best way to intuitively understand these mechanisms is to experiment with some of the computer simulations, changing parameters and conditions and observing the outcome. The program accompanying this book provides the tools required for this. For most of the simulations shown, prerecorded parameter files are available. The commands for starting the computer simulations are given in the figure captions. For instance, the command S23a and S23b will produce the simulations shown in Figure 2.3 (for details of the computer simulations see chapter 11).

2.3 Periodic patterns in space

If the range of the antagonist, i.e., the mean distance between the birth of a molecule and its decay, is smaller than the total field, several activator maxima will be formed at more or less regular distances. If we assume that the activator controls pigment production, stripes perpendicular to the growing edge will result, as seen on many shells. A simulation based on the activator-inhibitor scheme is shown in Figure 2.3. The inhibitor maxima are centered around the activator maxima, but have a more shallow distribution. Also, a possible reason for the irregularity in the spacing is clearly visible. Initially, some maxima develop too close together and don't survive the competition with the long-range inhibitor, leaving irregular-sized gaps in between. Since each maximum is surrounded by a cloud of inhibition, the maxima will not be shifted in order to form a more regular pattern.

Patterns in which particular structures are formed at more or less regular distances are common in many morphogenetic situations. The regular initiation of leaves on a growing shoot, the spacing of bristles, feathers and hairs may serve as examples. One case of a periodic pattern in a one-dimensional array of cells is the formation of heterocyst in the blue-green algae *Anabaena* (Wilcox *et al.*, 1973).

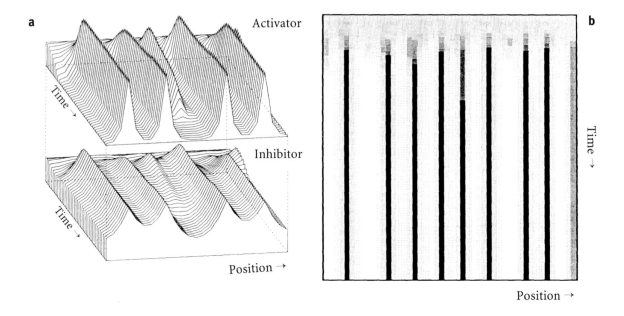

Figure 2.3. Stable periodic patterns in space. (a) Computer simulation in a linear array of cells. Concentrations of the activator (top) and the inhibitor (bottom) are plotted as functions of time. Since the size of the field is assumed to be much larger than the range of the inhibitor several maxima are formed. Due to the initiation by random fluctuations the spacing is somewhat irregular. Maxima can appear too close together and some of them will not survive the mutual inhibition. However, a maximum and minimum distance is maintained. The pattern is more regular if pattern formation is already at work during growth (see Figure 2.7). (b) A similar simulation with a space-time plot analogous to that on shells. (c) Shell of *Lyra taiwanica* [S23a, S23b].

Equation 2.1: The activator-inhibitor system

The following partial differential equations describe a possible interaction between the autocatalytic activator $a(x)$ and its antagonist, the inhibitor $b(x)$. They relate the concentration change per time unit of both substances as a function of the concentration present.

$$\frac{\partial a}{\partial t} = s\left(\frac{a^2}{b} + b_a\right) - r_a a + D_a \frac{\partial^2 a}{\partial x^2} \qquad (2.1.a)$$

$$\frac{\partial b}{\partial t} = sa^2 - r_b b + D_b \frac{\partial^2 b}{\partial x^2} + b_b \qquad (2.1.b)$$

where t is time, x is the spatial coordinate, D_a and D_b are the diffusion coefficients, and r_a and r_b the decay rates of a and b. A more detailed list of the individual terms will facilitate the reading of the equations.

sa^2/b The production rate. The activator a has a non-linear autocatalytic influence. For instance, two activator molecules must form a complex in order to satisfy the required non-linearity. The production is slowed down by the inhibitor b. The source density s describes the ability of the cells to perform the autocatalysis.

$-r_a a$ The rate of removal. The rate at which molecules disappear is, as a rule, proportional to the number of molecules present (like the number of individuals dying per year in a city is proportional to the number of inhabitants).

$D_a \partial^2 a/\partial x^2$ The exchange by diffusion. The exchange is proportional to the second derivative for the following reason. The net exchange of molecules by diffusion is, of course, zero if all cells have the same concentration. But the net exchange is also zero if a constant concentration difference exists between neighbouring cells, i.e. in the case of a linear concentration gradient. In this case each cell obtains the same amount of substance from its higher neighbour as it loses to its lower neighbour. In other words, it is not the change of concentrations but the change of concentration changes in space that is decisive in loss or gain by diffusion.

b_a Basic activator production. A small activator-independent activator production can initiate the system at low activator concentrations. It is required for pattern regeneration, for the insertion of new maxima during growth, or for sustained oscillations.

b_b Basic inhibitor production. A small activator-independent inhibitor production can cause a second homogeneous stable state at low activator concentrations. The system can be asleep until an external trigger occurs, for instance an influx of activator from a neighbouring activated cell. This term will play a role in the simulation of travelling waves (chapter 3).

Equation 2.2: Local instability and global stabilization - a simple calculation

A simple calculation should provide some intuition into why the interaction given in (2.1) can lead to stable patterns. For simplicity, we assume all constants are equal to 1, disregard diffusion and assume that even the inhibitor concentration is constant and equals 1. Equation 2.1a would be simplified to

$$\frac{\partial a}{\partial t} = a^2 - a \qquad (2.2a)$$

In this simplified version, the activator a has a steady state ($\partial a/\partial t = 0$) at $a = 1$. However, this steady state is unstable since for any concentration of a larger than 1, $a^2 - a$ will be positive and the concentration of a will further increase and *vice versa*. The reason for this instability lies in the over-exponential autocatalytic production in conjunction with a normal exponential decay. Now let us include the action of the inhibitor. Disregarding again any constants and diffusion, Equation 2.1b simplifies to

$$\frac{\partial b}{\partial t} = a^2 - b \qquad (2.2b)$$

which has a steady state at $b = a^2$. If we assume that the inhibitor reaches equilibrium relatively rapidly after a change in activator concentration, this change can be expressed as function of the activator concentration alone:

$$\frac{\partial a}{\partial t} = \frac{a^2}{b} - a \approx \frac{a^2}{a^2} - a = 1 - a \qquad (2.2c)$$

Therefore, if we include the action of the inhibitor, we obtain a steady state at $a = 1$ but this is stable since if a is larger than 1, $1 - a$ is negative and the concentration will return to the steady state of $a = 1$. To see why an interaction according to (2.1a,b) can generate a pattern we must take into consideration the fact that the inhibitor is assumed to diffuse much faster than the activator. Let us assume an array of cells in which all cells are at the steady state concentrations of a and b, except one cell that has a slightly increased activator concentration. It will also produce more of the inhibitor. However, after a small local perturbation the inhibitor concentration remains nearly constant since it diffuses rapidly (Figure 2.2). Thus, it is the average activator concentration that is decisive in inhibitor production. As mentioned, if the inhibitor remains constant, any deviation from the activator steady state will continue to grow since the steady state is unstable. However, after a substantial increase of the activator maximum, the inhibitor concentration can no longer be regarded as constant. As shown above, the action of the inhibitor leads to the stabilization of the autocatalysis. A new stable, steady state pattern will be reached. Thus, the formation of a stable pattern depends on local instability and global stabilization.

Equation 2.3: Saturation of autocatalysis

Saturation of activator production can be included in Equation 2.1a in the following way:

$$\frac{\partial a}{\partial t} = s \left(\frac{a^2}{b\,(1 + s_a a^2)} + b_a \right) - \cdots \qquad (2.3)$$

At low a-concentrations, $s_a a^2$ is negligible compared to 1 and the autocatalysis works as outlined above. With an increase in a, the term $s_a a^2$ becomes more dominant. Autocatalysis does not increase further since self-enhancement and slowing down due to saturation both become proportional to a^2.

When cell division causes the distance between two heterocyst cells to become more than about 7 cells, a new heterocyst is formed in between, corresponding to the insertion of a newly activated region during growth, as shown in Figure 2.7a.

2.4 The width of stripes and the role of saturation

Several factors determine the width of stripes, their distance apart, and the regularity of the distances. The range, i.e. the average distance that a molecule can travel in the time interval between its production and its disappearance depends on its diffusion rate as well as its half life. An increase in the diffusion rate of the activator leads to a broadening of the stripes, while an increase in the range of the inhibitor creates a larger region over which activated cells can suppress other cells from becoming activated. Thus a short range activator and a long range inhibitor would cause very narrow stripes at large distances. An example is given in Figure 2.4.

So far, the only limitations of autocatalysis that have been considered result from the action of the antagonist. It is conceivable, however, that other factors limit the maximum rate of autocatalysis too. For instance, an enzyme required for autocatalysis may be available only in limited amounts. At high activator concentrations the reaction would slow down since all available enzyme molecules would already be occupied. The consequence is a saturation of the autocatalysis since the maximum activator concentration reaches its upper bound (see Equation 2.3). Such saturation has severe influences on the outcoming pattern and a comparison between actual and simulated patterns indicates that saturation is frequently involved.

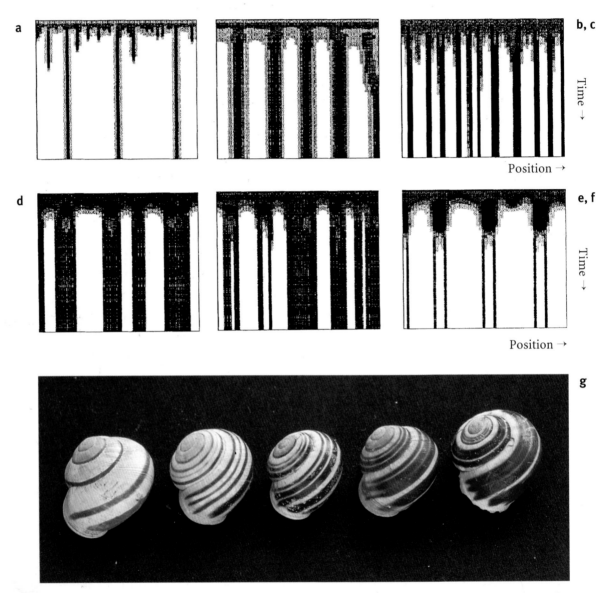

Figure 2.4. The width of stripes and the regularity of distances. (a) A short range activator and a long range inhibitor lead to narrow stripes at large distances. Since a maximum cannot be shifted, the distances are irregular. (b) Higher activator diffusion (D_a from 0.002 to 0.015) causes more regularly spaced broad stripes. (c) Saturation of the autocatalysis (s_a from 0.0 to 0.3) leads to broad stripes of irregular width and spacing. (d-f) Irregular stripes of different width and spacing will result if communication between the cells is switched off at an early stage. Pairs of stripes can arise (compare with Figure 2.1). (g) Stripes of different width on the garden snail *Cepea nemoralis* [S24a, S24d].

If the activator production has an upper bound due to saturation of autocatalysis, the activator concentration cannot surpass a certain level. Therefore, inhibitor production, and thus the inhibitory influence on neighbouring cells, is limited as well. The activated region will increase in size until sufficient inhibitor is produced. In large activated regions, the accumulation of inhibitor at the centre causes near-instability and the tendency to decay into two smaller maxima. The maxima can be broad although the range of the activator is small. Activator diffusion determines the minimum width of a stripe only. The individual maxima can have different widths. The transition between activated and non-activated areas can be sharp since diffusion of the activator may be slow. Due to saturation, the ratio between activated and non-activated cells becomes independent of the size of the field since, for instance, if less than the normal number of cells are activated, the inhibitor concentration will be too low and the activated regions will enlarge until the correct ratio is established. Thus, if activator production saturates, a pattern of broad but irregularly spaced stripes is expected. An example is given in Figure 2.4. With increasing saturation autocatalysis becomes less efficient until pattern formation is no longer possible. Therefore, the activated region cannot become much larger than the non-activated region due to saturation. If the pattern consists of narrow non-pigmented lines in a pigmented background, other mechanisms are expected to be involved (see Figure 2.8).

A further consequence of the saturation of autocatalysis is that a maximum can be shifted relatively easily to a more favourable location. If a portion of the maximum is disfavoured, for instance by another activated region in the vicinity, this part of the maximum can become deactivated in favour of a portion that is not subject to this inhibition. An example is provided in connection with an increase in field size due to growth (Figure 2.6).

2.5 Early fixation of a pattern

The stripes on the shells shown in Figure 2.4 have very different widths at irregular distances although they belong to the same species and were found in the same place. This phenomenon could be based on the early fixation of an evolving pattern. Imagine that at an early state the exchange of substances by diffusion ceased, for instance, by closing the gap junctions between the cells. Usually, the non-activated cells are no longer inhibited and return to the homogeneous steady state activation. If, however, the inhibitor has an additional small activator-independent production term (b_b in Equation 2.1b), a second stable steady state exists at low activator concentrations. All cells above a certain threshold adopt the high steady state concentration, while the remaining cells adopt the low. The activation of a cell will be faithfully transmitted to the progeny of each cell. During early phases of pattern formation, the pattern is much less regular and depends on accidental initial conditions (see Figure 2.3). If this pattern is frozen by the mechanism

Equation 2.4: The activator-depleted substrate system

An antagonistic effect can result from the depletion of the substrate $b(x)$ needed for autocatalysis. The interaction may have the following form

$$\frac{\partial a}{\partial t} = sba^{*2} - r_a a + D_a \frac{\partial^2 a}{\partial x^2} \tag{2.4.a}$$

$$\frac{\partial b}{\partial t} = b_b(x) - sba^{*2} - r_b b + D_b \frac{\partial^2 b}{\partial x^2} \tag{2.4.b}$$

Usually this interaction is used with a saturation term s_a and

$$a^{*2} = \frac{a^2}{1 + s_a a^2} + b_a$$

The following is a description of some of the terms used in these equations:

sba^{*2}　The production rate of the activator. The non-linear autocatalysis is proportional to the substrate concentration b. This production leads to a decrease in b by the same rate $-sba^{*2}$.

b_a　Basic activator production. A small activator-independent production can initiate the system at low activator concentrations. It is required for pattern regeneration or for sustained oscillations.

$b_b(x)$　Production rate of the substrate b. Usually this is the same in all cells. However, for the patterns discussed in chapter 4, different rates along the growing edge play a crucial role.

$-r_b b$　A destruction term independent of removal due to autocatalysis. It is not necessary for pattern formation. It limits the maximum substrate concentration in non-activated regions. If significant, a spontaneous trigger at low activator concentration may no longer be possible, similar to the effect of the term b_b in Equation 2.1b. It plays an important role in the transition between an excitable system and sustained oscillations (see Figure 3.3).

$s_a a^2$　This saturation term has similar consequences for the activator-inhibitor system as the one discussed in Equation 2.3: maxima become broader and the number of activated cells are a fraction of the total number of cells. For computational purposes, a small saturation term is also helpful in this interaction to avoid numerical instabilities.

outlined above, it will be much less regular. Figure 2.4d-f show some simulations. Necessarily connected with such mechanisms is the loss of regulatory properties. For instance, the insertion of new maxima during growth is no longer possible. An example is given in Figure 2.6.

A further effect of ceasing diffusion may be the generation of pairs of stripes. Regions with high activator concentration also have high inhibitor concentrations. After an abrupt termination of diffusion, the centre of a maximum may become deactivated. Activation survives only in the two marginal regions (Figure 2.4f). A natural example of such a pairwise arrangement of narrow stripes was given in Figure 2.1.

2.6 The activator-depleted substrate scheme

As mentioned earlier in connection with the sand dune example, the antagonistic effect can also result from the depletion of the substrate $b(x)$ that is consumed during production of the autocatalytic activator $a(x)$. A possible interaction is given in Equation 2.4. It is similar to the Brusselator reaction proposed by Prigogine and Lefever (1968) but is somewhat simpler. In order to allow stable pattern formation, sufficient substrate must be supplied to maintain a steady activator production, i.e. $b_b > r_a$. Again, the diffusion of the substrate must be much faster than that of the activator, i.e., $D_b \gg D_a$.

The formation of a periodic pattern by an activator-depleted substrate model is shown in Figure 2.5. For demonstration purposes, rather than random fluctuations in the source density s, a somewhat elevated activator concentration in the leftmost cell is assumed. This cell forms a full maximum, causing a depression in the substrate concentration. A second maximum can develop only at a distance, and so on. The patterning spreads like a wave and the spacing is regular.

As shown below, the activator-inhibitor and the activator-depleted substrate models lead to different patterns in some situations (see Figure 2.7). Although the two mechanisms appear so different in their molecular requirements, they may have ultimately the same basis. The interactions described by Equations 2.1 and 2.4 are idealized. In the activator-inhibitor scheme it is assumed that the activator is produced from a pool of precursor molecules that is infinitely large. Its depletion is assumed to be negligible due to the antagonistic action of the inhibitor. In the depletion scheme, on the other hand, the activator becomes degraded. It is conceivable that a degradation product acts as inhibitor in the autocatalysis due to competitive inhibition with the intact activator molecules. Thus, only minor evolutionary modifications may be required for a change from one regime to the other. We will use both schemes as well as combinations.

The activator-depletion mechanism has an inherent limit of maximum activator production since the activator production comes to rest if sufficient substrate is no longer available. Therefore, patterns generated using this mechanism show

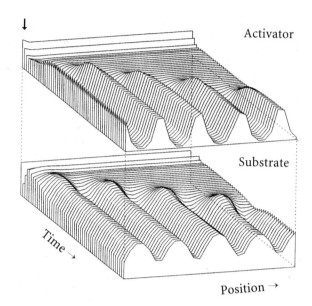

Figure 2.5. Pattern formation based on an activator-depleted substrate scheme. Initiation by a locally elevated activator concentration (arrow). This elevation grows to a full maximum at the expense of the substrate in the surroundings. A further maximum can be formed only at a distance. For demonstration purposes, the general activator concentration is higher than the steady state. A rapid regulation back to the steady state occurs [S25].

relatively broad maxima that tend to shift into regions in which high substrate concentrations are still available. Therefore, the spacing is generally more regular. The width of the maxima is of the same order of magnitude as the space that separates them. In contrast, in the activator-inhibitor system the activated region may be only a small portion of the total field (see Figure 2.7).

2.7 The influence of growth

In the generation of many shell patterns the elongation of the margin due to growth cannot be neglected. By inserting new cells, the distance between pigment producing cells enlarges. This can either cause the insertion of new pigmentation lines or a broadening and branching of existing lines. In all such cases, the ratio of pigmented to non-pigmented regions remains more or less constant. Examples are given in Figure 2.6.

Insertion of new lines indicates the formation of a new activated region between existing maxima. Due to growth, the distance between the activated regions increases too. In the activator-inhibitor scheme large distances between existing maxima cause the inhibitor concentration to become so low that a small activator-independent activator production (b_a, Equation 2.1a) is sufficient for the onset of autocatalysis (Figure 2.7).

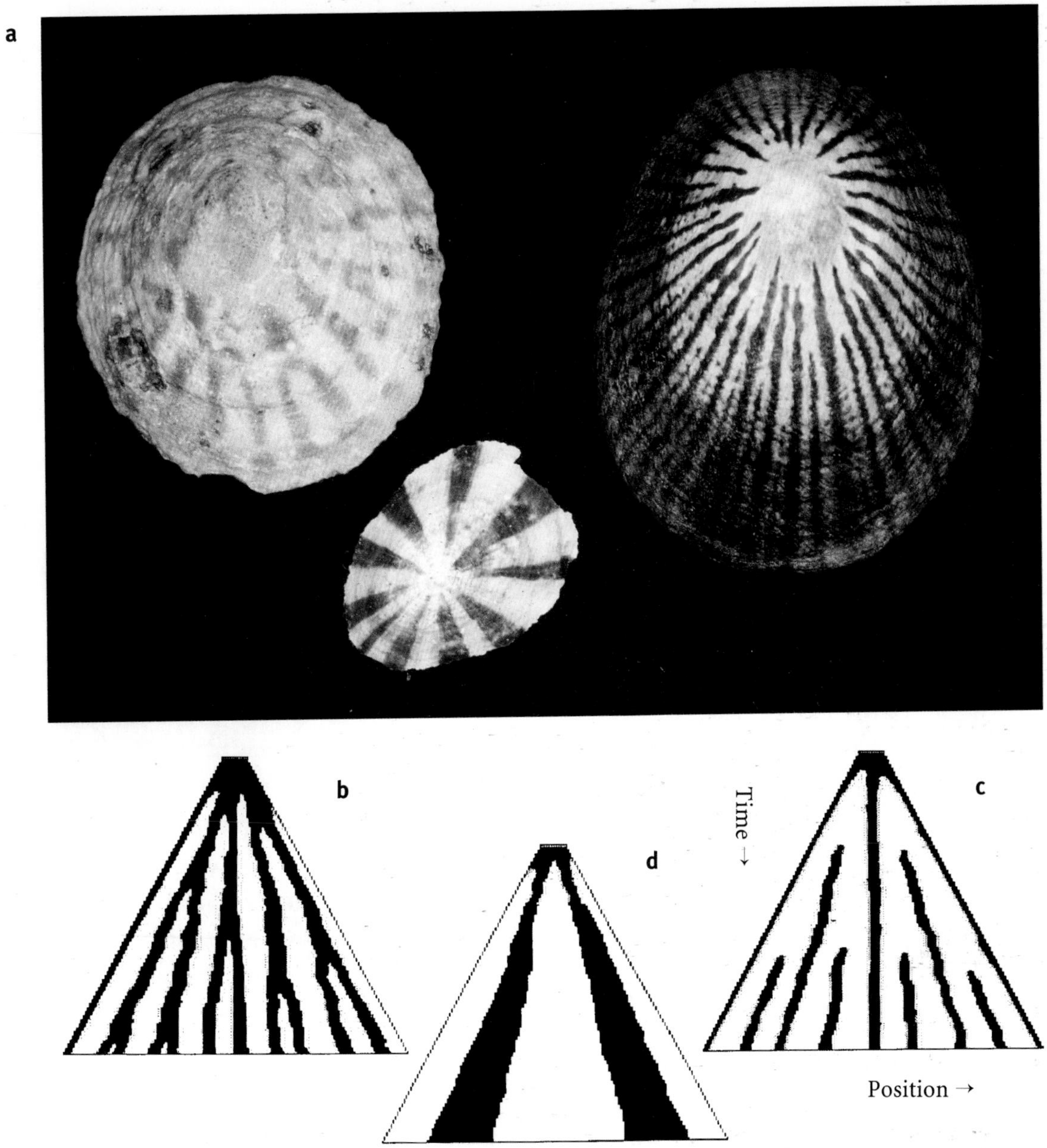

Figure 2.6. Pattern regulation due to growth. (a) Natural patterns. Bifurcation of existing lines, insertion of new lines, or a wedge-like widening of existing lines occurs. (b)-(d) Computer simulations of an activator-inhibitor model. Growth is simulated by the insertion of two new cells, one in each half after a certain time interval. (b) If activator production saturates, the broader maxima become still broader until a split and shift occurs. (c) New maxima become inserted if the distance between existing maxima becomes too large. (d) Wedge-like patterns result if once-activated cells propagate pigment production in a clonal way. In this simulation it is assumed that the activator-inhibitor system is bistable. After an early cessation of diffusion, all cells activated above a certain threshold remain activated while the remaining cells become completely non-activated [S26b, S26d].

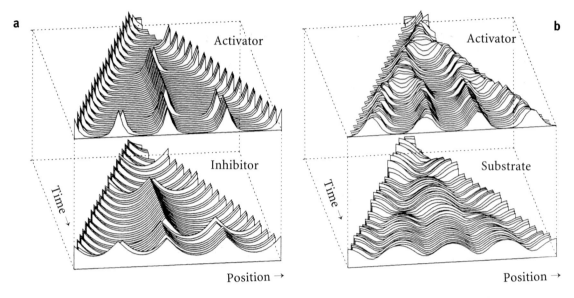

Figure 2.7. Different behaviour during growth. (a) With the activator-inhibitor mechanism, new regions become activated if the inhibitor becomes too low to suppress the onset of autocatalysis in the enlarging space between the maxima. (b) By contrast, in the activator-substrate model, a tendency exists to shift existing maxima towards higher substrate concentrations. This shift may be connected with a split of a maximum. With saturation, an activator-inhibitor system would behave in a similar way (see Figure 2.6) [S27a, S27b].

As mentioned above, saturation of autocatalysis leads to a broadening of the stripes. A maximum that is too broad may become deactivated in the centre, causing a split in a line. Thus, the shift and split of lines indicate either an activator-inhibitor scheme with saturation (Figure 2.6) or a depletion mechanism (Figure 2.7b) with its inherent saturation. Wedge-like structures emerge if a system is tuned to bistability and the communication by diffusion is switched off at an early stage since the state of activation is kept in the progeny of each cell.

2.8 Inhibition via destruction of the activator

An antagonistic reaction can result not only from the slowing down of the activator production rate, but also from an increase in its destruction rate. Turing (1952) used such a mechanism in his pioneering paper. An example is given in Equation 2.5. Segel and Jackson (1972) have proposed such a mechanism to describe the dynamics and pattern formation of two spreading populations in which one species acts as prey.

Such interaction has several drawbacks in normal pattern formation. If the inhibition results from increased destruction rather than reduced production, much energy would be required for the high turn-over of molecules. Moreover, the range of the activator, i.e., the mean distance between the production and the decay of

Equation 2.5: Inhibition via activator destruction

As an alternative to slowing down activator production, the inhibitor may also accelerate activator removal:

$$\frac{\partial a}{\partial t} = s(a^2 + b_a) - r_a ba + D_a \frac{\partial^2 a}{\partial x^2}$$
(2.5.a)

$$\frac{\partial b}{\partial t} = sa^2 - r_b b + D_b \frac{\partial^2 b}{\partial x^2} + b_b$$
(2.5.b)

sa^2 Production rate. In contrast to Equation 2.1.a, it is independent of inhibitor concentration.

$-r_a ba$ Rate of activator removal is proportional to the number of activator *and* inhibitor molecules.

the activator, changes with the formation of a local activator maximum, since the inhibitor also increases and the lifetime of the activator becomes reduced. Therefore, the width of the peaks would shrink. As shown in the next chapter, a shortening of the activator lifetime may cause a transition to oscillating patterns. This is certainly not desirable for pattern formation during normal embryogenesis. But in shell patterning where oscillating patterns play an essential role, this mode cannot be ruled out.

2.9 Autocatalysis by an inhibition of an inhibition

The activator-inhibitor mechanism as given in Equation 2.1a,b is, of course, only one example of the many possible molecular realizations that satisfy the general requirements of the theory. Pattern formation does not require a molecule with direct autocatalytic regulation. Autocatalysis can be a property of the system as a whole. For instance, if two substances, a and c exist and a inhibits c and *vice versa*, a small increase in a above an equilibrium leads to a stronger repression of c production by a. This, in turn, leads to a further increase of a, in the same way as a would increase if it were autocatalytic. The same holds for c. a and c together form a switching system in which either a or c is high. The switch of the λ phage between the lytic and lysogenic phase is based on such an inhibition of an inhibition (Ptashne et al., 1980). To allow pattern formation, a long range signal is required that interferes with the mutual competition, i.e., with the indirect self-enhancement of one or the other substance. For instance, if a has won the a-c competition in a particular region, c must win in the surrounding regions.

Equation 2.6: Indirect autocatalysis by an inhibition of an inhibition

Autocatalysis may result from the interaction of several molecules. In this case it results from an inhibition of an inhibition

$$\frac{\partial a}{\partial t} = \frac{s}{s_a + c^2} - r_a a + D_a \frac{\partial^2 a}{\partial x^2} + b_a \tag{2.6.a}$$

$$\frac{\partial b}{\partial t} = r_b a - r_b b + D_b \frac{\partial^2 b}{\partial x^2} \tag{2.6.b}$$

$$\frac{\partial c}{\partial t} = \frac{s}{s_c + a^2/b^2} - r_c c + D_c \frac{\partial^2 c}{\partial x^2} \tag{2.6.c}$$

a, c Two substances that inhibit each other's production. Together they form a switching system in which one of the substances becomes fully activated.

b The rapidly diffusing substance required for pattern formation. It undermines the inhibition of c production by a molecules and therefore acts as an inhibitor.

s_a, s_c These constants determine the lowest level of a and c concentration. In this way they also determine the maximum concentration a substance may reach and function similar to the saturation term s_a in Equation 2.3. A more general discussion of possible mechanisms and their equivalences is provided by Gierer (1981).

A possible realization would be that the a molecules control the production of a substance b which, in turn, either inhibits a or promotes c production. These modes are equivalent since self-limitation in competing systems is equivalent to support for the competitor. In Equation 2.6 an interaction is described in which the diffusible antagonistic substance b is produced under control of a molecules and undermines the repression of c production by the a molecules. The b molecules may be a decay product of the a molecules. No direct autocatalytic interaction is assumed.

As mentioned in the introduction, similar patterns may be formed with reverse pigmentation. In the mechanism just described, a and c complement each other. In some molluscs a, while in in others c may be used as a signal to induce pigmentation. With an otherwise unchanged mechanism, a pigmented pattern on a non-pigmented background would be formed, or *vice versa*. An example is given in Figure 2.8.

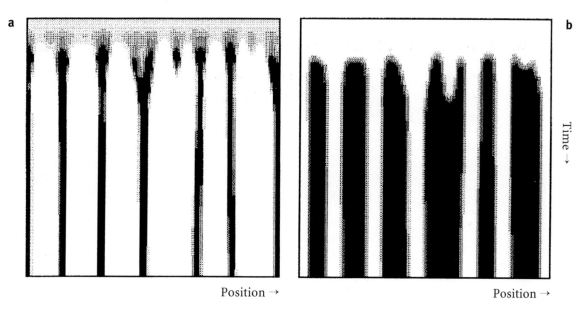

Position →   Position →

Figure 2.8. White pattern on a dark background. Two complementary distributions result if the autocatalysis is caused by the mutual inhibition of two substances (*a* and *c* in Equation 2.6). Depending on the substance that is used to trigger pigmentation, the pigmented region can be larger than the non-pigmented one [S28].

2.10 Formation of graded concentration profiles

The growing edge of a shell has been treated as a line and all cells along this line as equivalent. However, shells of snails are strongly asymmetric. One side usually forms the cone while the other side forms a finer tip and the opening for the syphon. The pattern formed along the edge frequently shows systematic non-uniformities, for example, two bands of stronger pigmentation at particular positions (see Figure 1.12). The cells, therefore, require some information about their position along the growing edge, where the shoulder or stronger pigmentation must be formed. This is a very general problem in early embryogenesis where primary body axes must be established to determine the positions of the head and tail.

A property of this positional mechanism is that polar concentration profiles will be formed if the range of the activator is comparable with the size of the field. A critical size must be exceeded in order for pattern formation to be possible since a rapid redistribution of the activator within a small field would wipe out any pattern. If this critical size is reached, only a polar pattern can be formed since a central maximum would require space for two slopes and this space is not available at the critical size. Therefore, a high concentration would appear on one side and a low concentration at maximum distance on the other (Figure 2.9). The local concentration of the activator and/or the inhibitor can provide positional

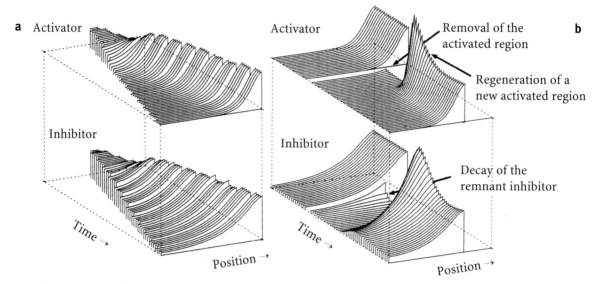

Figure 2.9. Generation of a polar pattern and its regeneration. (a) After the field has grown to a size comparable to the range of the activator, small random fluctuations in the cells performing the autocatalysis (the source density s in Equation 2.1) are sufficient to initiate pattern formation. At this critical size only one marginal maximum can be formed. The result is a polar pattern that can be used as positional information. This pattern has been calculated with an activator-inhibitor mechanism. An activator-substrate model is less convenient for generating gradients due to the tendency of the maximum to shift towards higher substrate concentrations (see Figure 2.7). (b) After removal of the activated region, a new activator maximum can be formed once the remnant inhibitor has decayed [S29].

information (Wolpert, 1969) about where a cell is located. A graded profile, once formed, can be stable if the formation of secondary maxima are suppressed. This is the case, for instance, when the basic activator production b_a is low enough and/or a basic inhibitor production b_b exists (see Equation 2.1; compare Figure 2.9 with Figure 2.7).

Since a homogeneous activator-inhibitor distribution represents an unstable situation, an inhomogeneity would initiate pattern formation. Any asymmetry imposed by a mother onto an egg leads to a predictable and inheritable orientation of this polar distribution. According to the model, such stimulus or asymmetry only orients the pattern. Due to self-regulation the resulting pattern is fairly independent of the initial stimulus. Therefore, no precision is required in the triggering signal. A stronger initiating asymmetry has the advantage that the pattern reaches the final steady state much faster since no time-consuming competition is required between opposite sides of the field to determine which side will win. Moreover, in a somewhat larger field the danger of forming a symmetric rather than a polar pattern is lower. The asymmetry can, but need not be, specific. Any asymmetry in oxygen supply, in pH or in temperature may be sufficient. Local disadvantages cause the formation of the maximum at the opposite side. Pattern initiation may result from fluctuations in activator concentration or from an inhomogeneity in

Figure 2.10. Examples of shell patterns generated by a two-dimensional process. The molluscs engulf their shells with two ectodermal protrusions and impose a copy of the skin patterns by secretion of pigment onto the shell. The line where the two protrusions contact each other generates the seam-like structure on the top of both shells (*Cypraea scurra* and at right *Cypraea tigris*).

the source density (*s* in Equation 2.1), i.e., from the ability of the cells to perform the pattern forming reaction.

An activator-inhibitor system exhibits substantial pattern regulation. By removing the site of high activator concentration, the site of inhibitor production is removed too. The remnant inhibitor decays until a new activator maximum is triggered in the remaining cells due to the low level of activator production (b_a in Equation 2.1). The pattern is restored in a self-regulatory way (Figure 2.9).

Diffusion can generate a graded distribution of substances in a reasonable amount of time (1-2 hours) only if the field of cells is smaller than about 2mm (Crick, 1970). Wolpert (1969) has pointed out that at the stage at which pattern formation takes place all known biological systems are small, less than about 100 cells and 1mm across. For fields that grow to a larger size it is expected that the concentration pattern becomes translated into a pattern of position-dependent stable gene activation. The evoking signal, the graded distribution of the "morphogen", is then no longer required. Corresponding models based on self-enhancement and competition of genes have been proposed and have found much experimental support (Meinhardt, 1978, 1992).

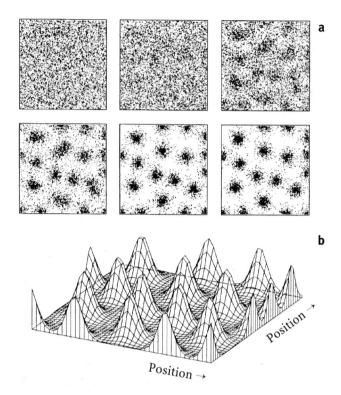

Figure 2.11. Pattern formation in two dimensions. If the range of the inhibitor is much smaller than the total field, a somewhat irregular pattern of patches emerges, but a maximum and minimum distance is maintained. (a) Stages. (b) Final distribution.

2.11   Pattern formation in two dimensions

Although pattern formation on shells can generally be regarded as a one-dimensional process that keeps a time record in the second dimension, some shells become decorated by an actual two-dimensional process (Figure 2.10). The mechanism outlined above produces activated regions at more or less regular distances in two dimensions. Figure 2.11 shows the emergence of such a pattern from random fluctuations.

As previously mentioned, saturation leads to a broadening of the maxima. In two-dimensional pattern formation this has a non-trivial effect: the formation of stripes. Stripes are very common in embryonic development. These stripes should not be confused with the stripes generated by the one-dimensional processes discussed earlier. Stripe formation in two dimensions is achieved by an activated region that has a large extension in one dimension and a small extension perpendicular to it. How is this possible? In the model, due to saturation activated cells have to tolerate other activated cells in their vicinity. If the activator shows modest diffusion, activated regions tend to occur in larger coherent patches since the probability is high that an activated cell's neighbour will become activated

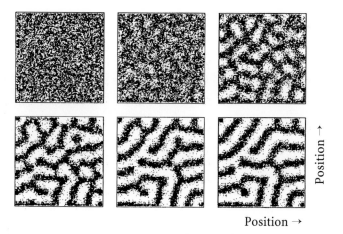

Position →

Position →

Figure 2.12. Stages in the formation of stripes. If the autocatalytic reaction saturates at low activator concentrations, more cells remain activated but at a lower level. Stripes are the preferred pattern since activated cells have activated cells as neighbours and non-activated cells are nearby into which the inhibitor can diffuse.

too. On the other hand, activated cells must be close to non-activated cells into which the inhibitor can diffuse or from which substrate can be obtained. These two seemingly contradictory requirements, large coherent patches and proximity to non-activated cells are satisfied if a stripe-like pattern is formed (Figure 2.12). Each activated cell is bordered by other activated cells but is also close to non-activated cells. If initiated by random fluctuations the stripes will also have random orientation and will exhibit some bending and bifurcations.

In conclusion, relatively simple molecular interactions based on local self-enhancement and long range inhibition allow the formation of stable patterns starting from originally more or less homogeneous conditions. These patterns can be used to create particular structures at particular positions during development, for instance, to establish embryonic axes or to initiate periodic structures in space. The formation of pigmentation stripes on the shells of molluscs that are oriented parallel to the direction of growth is only a very special case of this type of interaction in which pattern formation plays a central role.

Figure 3.1. Parallel and oblique lines, the traces of synchronous oscillations and travelling waves. On the shell of *Amoria ellioti* (top), the stripes are more or less perpendicular to the direction of growth. They result from an almost synchronous alternation between pigment producing and non-producing phases. It is remarkable that the distance between the lines remains constant in regions of different shell diameter. Although the lower left pattern (*Nerita communis*) looks similar on the picture, it has a different origin. From the lower right figure it becomes clear that the stripes are oriented oblique to the direction of growth. Therefore, a pigmented region has triggered a neighbouring region and so on, forming travelling waves at regular intervals. Minute irregularities and finer changes in the pigmentation indicate the orientation of the growing edge at the corresponding stage.

Chapter 3

Oscillations and travelling waves

A very important class of shell patterning is caused by pigment productions that occur only during a short time interval, followed by an inactive period without pigment production. Stripes parallel to the growing edge and oblique lines belong to this class of patterns. Oscillations can occur if the antagonist reacts too slowly. A change in activator concentration cannot be immediately regulated again causing activation to proceed in a burst-like manner. Only after a sufficient accumulation of the inhibitor, or after a severe depletion of the substrate, will activator production collapse. A refractory period will follow with very low activator production in which either the excess inhibitor will degrade or the substrate will accumulate until a new activation becomes possible. The condition for oscillatory activations is the reverse of that given for stable patterns. In an activator-inhibitor scheme, oscillations occur if the decay rate of the inhibitor is smaller than that of the activator i.e., if the condition $r_b < r_a$ in Equation 2.1 (page 23) is satisfied.

In the activator-depleted substrate model, oscillations occur if the rate of substrate production is too low to maintain activator production in a steady state, i.e., if $b_b < r_a$ in Equation 2.4 (page 28). The onset of oscillations in a depletion-driven system can be observed using an everyday experiment. If a thick candle burns for a while a deep hole will form in the wax and the flame will begin to flicker. The reason is that the oxygen supply becomes insufficient for a large flame. The flame shrinks consuming less oxygen, thereby allowing the oxygen concentration to recover and the flame becomes larger again; and so on.

Depending on the parameter, this reaction can exhibit different behaviours. A cell may become activated with an internal periodicity. Or, it may be arrested in an excited state. Small changes in a parameter can lead to a transition from one to the other mode. Figure 3.2a, for instance, shows the switch between oscillations and stable activation occurring in a single cell and caused by a change in the decay rate of the activator. Theories on oscillations and waves in excitable media are well developed (see, for instance, Prigogine and Lefever, 1968; Winfree, 1980; Segel, 1984; Glass and Mackey, 1988; Murray, 1989). The different modes will be discussed here in some detail since they are the tools for deducing the parameters required for the patterns observed on shells.

Equation 3.1: Finite activator production at low inhibitor concentration: Michaelis-Menten kinetics

At very low inhibitor concentrations, inhibitor production must remain finite. This is accomplished using the Michaelis-Menten term s_b. Together with the saturation term s_a introduced in Equation 2.3, Equation 2.1a (page 23) would obtain the following form:

$$\frac{\partial a}{\partial t} = s \left(\frac{a^2}{(s_b + b)\,(1 + s_a a^2)} + b_a \right) - r_a a + D_a \frac{\partial^2 a}{\partial x^2}$$

s_b has an effect similar to the non-zero baseline inhibitor production b_b in Equation 2.1b. Both effects limit activator production at low inhibitor concentrations. This limitation can lead to a second stable state at low activator concentrations.

In an oscillating activator-inhibitor mechanism, inhibitor concentrations can become very low. So far, it has been assumed that the rate of activator production is inversely proportional to inhibitor concentration. This is a simplification since, of course, the production rate remains finite even if no inhibitor is present. At very low concentrations, a Michaelis-Menten term s_b, as introduced in Equation 3.1, must be considered.

If either the Michaelis-Menten constant s_b or the basic inhibitor production b_b is high, but the basic activator production b_a is low, a spontaneous trigger may not occur since the effective inhibition remains too high. Such a system will no longer oscillate but will remain arrested in an excited state. The external addition of small amounts of activator can initiate either a single burst (Figure 3.2b) or a sequence of bursts, depending on the duration of the external activator supply. As shown below, this mode plays a crucial role in the formation of travelling waves.

A basic activator production b_a above the threshold leads to sustained oscillations. Whenever the inhibitor drops below a critical value, a new activator burst is triggered. A higher baseline activator production enables an earlier onset of the self-enhancing process. In this earlier phase inhibitor concentration is still higher. Thus, the bursting proceeds less dramatically (Figure 3.2c). This effect is well-known in economics as a means of dampening oscillations. By government investments in periods of economic depression, i.e., by artificial inputs into a process that hopefully becomes self-enhancing, attempts are made to make the depression less severe and to initiate the next upswing earlier. That the following boom is less exaggerated is another helpful side-effect of this strategy.

The same type of behaviour may occur in an activator-substrate scheme (Figure 3.3). Substrate removal independent of consumption by the autocatalytic pro-

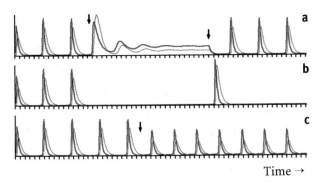

Figure 3.2. Oscillations in an activator - inhibitor system. Calculated on a single cell, concentrations of the activator (green) and inhibitor (red) are plotted as a function of time. Oscillations occur if the inhibitor requires longer than the activator to reach an equilibrium. (a) A temporary reduction of the decay rate of the activator ($r_a = 0.08 \rightarrow 0.03$) can lead to a transition to a stable situation. (b) If the basic activator production is too low and either the basic inhibitor production b_b or the Michaelis-Menten constant s_b is sufficiently large, the autocatalytic cycle cannot be fired at low inhibitor concentrations. Oscillation stops ($b_a = 0.04 \rightarrow 0.01$) and the system remains in an excited state. A short, small increase in an external activator supply ($b_a = 0.01 \rightarrow 0.04$) can cause a single burst. (c) A higher basic activator production leads to smaller peaks in more rapid succession ($b_a = 0.04 \rightarrow 0.08$). Each sequence (a-c) begins with the same standard conditions. The time of the parameter change is indicated by an arrow [GT32].

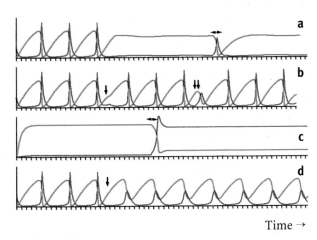

Figure 3.3. Oscillation in an activator-substrate scheme. (a) If the substrate supply is too slow, the activator concentration (green) oscillates. The substrate concentration (red) collapses during activation. After sufficient accumulation of substrate, a new pulse is fired. A loss of substrate, independent of activator production ($r_b = 0 \rightarrow 0.008$), can reduce the maximum substrate concentration. Oscillation may stop but the system remains excitable. A small external addition of activator for a short time interval ($b_a = 0.02 \rightarrow 0.06$) can initiate a single activation cycle. Due to the reduced substrate concentration, the pulse is somewhat smaller. (b) The result of an external activator pulse on an oscillating system depends on the phase. If the substrate concentration is low (arrow) no peak can be triggered. The additional substrate consumption may even delay the next peak. If activator addition occurs at a later phase, a premature peak with a lower amplitude (double arrow) will result. (c) The system becomes bistable if the substrate production is high enough to maintain a steady state activation but the loss of substrate via r_b suppresses the onset of a spontaneous activation. A short temporary increase in the external activator supply (double arrow) causes a permanent transition from low to high activation (and from high to low substrate concentration). (d) Saturation of the activator production ($s_a = 0 \rightarrow 1.0$) limits the height of the peak but elongates the duration of the pulses [GT33].

cess ($r_b > 0$ in Equation 2.4b) limits the maximum concentration that the substrate may reach. This level may become insufficient to release a spontaneous trigger of activation. The sustained oscillations are suspended but the system remains in an excited state similar to that accomplished by s_b or b_b in the activator-inhibitor system. Again, whether oscillations take place also depends on the basic activator production b_a required for the initiation of each subsequent burst. Depending on the phase of the oscillatory cycle, a short pulse of externally added activator can either advance or delay the next burst (Figure 3.3b). If the time constants allow a stable activation but the maximum substrate concentration is insufficient for a trigger the system becomes bistable. A short pulse of additional activator can accomplish a permanent transition from low to high activation (Figure 3.3c). Thus, small differences in parameters can cause very different behaviour within the system. The effects of these differences are basic to the understanding of different shell patterns.

3.1   The coupling between the oscillators by diffusion

In the overall pattern that emerges on shells, the coupling between oscillators is decisive. A strong coupling synchronizes the oscillations and the resulting patterns are stripes parallel to the growing edge (Figure 3.4). Since the total length of the edge is very long, even small phase differences between neighbouring oscillators can accumulate such that distant regions are out of phase. At a given time pigment may be produced in some positions but not at remote locations. An example was given in Figure 1.4. This indicates that the synchronization of oscillations does not result from external influences, such as the seasonal changes that effect tree rings, since these would cause strict synchronization. The synchronization may fail between regions of larger and smaller circumferences on shells and lines may end blindly (Figure 3.4c). This phenomenon will be examined again later (Figure 4.14).

Synchronization can result not only from diffusion of the activator but also from diffusion of the antagonist. For instance, a cell that is activated somewhat later than its neighbours delays the neighbouring cells *via* its later emitted inhibitor, thus enforcing synchronization. Therefore, a mere change in inhibitor lifetime may cause a transition from a stable pattern in time to synchronous oscillations. On shells, this corresponds to a transition from stripes perpendicular to the growing edge to parallel stripes (Figure 3.5c). Related transitions can be induced in shells. In a river polluted by salt containing waste water from potassium mining, Neumann (1959a,b) observed corresponding changes on the shells of the freshwater snail *Theodoxus fluviatilis* L as a function of the salt content. This change can be reproduced in the laboratory (see Figure 1.11).

Very frequently shell patterns exhibit rows of dots (see Figure 4.13). It would appear straightforward to assume that these patterns result from a combination of the lateral inhibition scheme ($D_b \gg D_a$) localizing activations, and the oscillation

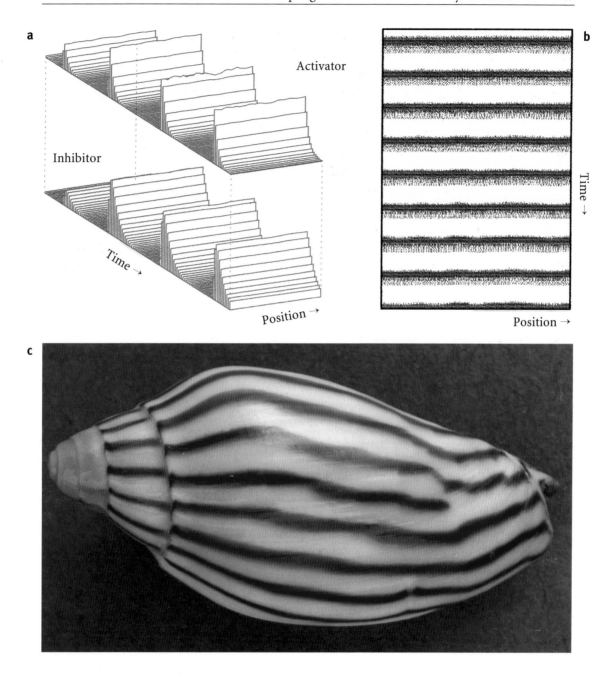

Figure 3.4. Stripes parallel to the direction of growth (perpendicular to the growing edge). This pattern indicates a nearly synchronous oscillation of pigment production. (a) Pattern formation by the interaction of an autocatalytic activator (top) and its antagonist, an inhibitor (bottom). Oscillations occur since the inhibitor has a longer time constant (lower decay rate) than the activator. High diffusion of the activator enforces a near-synchronization between neighbouring cells. Nevertheless, a substantial phase difference can accumulate between distant cells. (b) The resulting pattern in a space-time plot analogous to that formed on shells. Activator concentration is indicated by the density of the dots. (c) Pattern on the shell of *Amoria dampieria*; [S34].

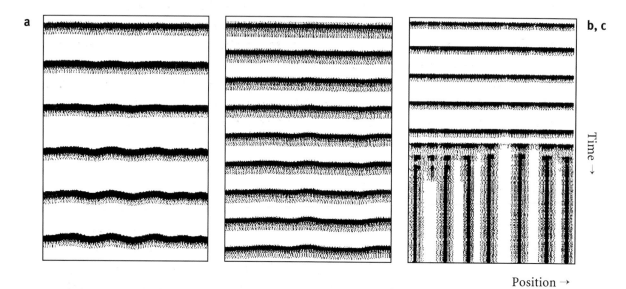

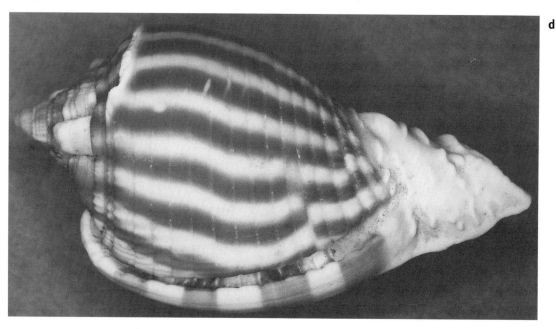

Figure 3.5. The width of stripes. (a) without saturation, a high activator concentration is present only for a short time interval. (b) If the autocatalytic process saturates ($s_a > 0$ in Equation 2.3), the width of the stripes becomes broader in relation to the interstripes. (c) A rapid diffusion of the inhibitor is insufficient for the formation of a pattern in space. However, a shortening of the inhibitor life time is sufficient for a transition from a homogeneous oscillation to a stable pattern in space (see the experimental observation in Figure 1.11). This leads to a switch between stripes parallel and perpendicular to the growing edge. (d) *Phalium striatum*, a shell with stripes and interstripes of about the same width. Note that at the bulge the pigment forms a pattern in space [S35a, S35b, S35c].

scheme ($r_b < r_a$) enforcing periodic activations. However, lateral inhibition requires time for one region to suppress surrounding regions (see Figure 2.3). This time is not available if the activation proceeds in a burst-like manner. As shown earlier, the diffusion of the antagonist leads to synchronous bursts of activation. Before the excess of inhibitor produced in one region can reach the neighbouring region *via* diffusion, full activation has already occurred there that will cause a down-regulation (Figure 3.5). As will be shown below, rows of dots require a superposition of two pattern forming systems (see Figure 4.11 and 4.13).

3.2 The width of bands and interbands

On many shells with parallel stripes, the dark stripes have approximately the same width as the lighter regions in between. In the simplest model, however, the activation is a short burst followed by a long period without activation (Figure 3.5a). This changes if the autocatalysis saturates, a process introduced earlier to model the width of stable stripes (see Figure 2.4). By saturation, the maximum activator concentration is limited and it requires more time either to accumulate sufficient inhibitor or to consume all available substrate. Therefore, it takes longer for the activation to cease. Figure 3.5a,b provide simulations with and without saturation.

However, the actual mechanism by which broad stripes are generated may be more complex. The natural patterns show bands of a nearly constant density of pigmentation. In contrast, simulations involving saturation of the autocatalysis produce a slow increase followed by a slow decrease of activation (see also Figure 3.3d). One possibility may be that pigmentation depends in a non-linear way on exceeding a certain threshold of activator concentration. It will be shown in chapter 7, however, that more complex patterns suggest the switching on and off of pigmentation are independent processes. This separation leads to periods of constant pigmentation density between the two signals (see Figure 7.5).

3.3 Oblique lines: travelling waves in an excitable medium

In many species pigmentation lines oblique to the growing edge are formed. An example is given in Figure 3.6. As mentioned, this type of pattern can be regarded as a time record of travelling waves of pigment production along the pigment-producing mantle gland at the growing edge. A cell with a high activator concentration "infects" its neighbour causing, with some delay, a burst of activation in this cell also, and so on. The chain of reaction can proceed only in one direction since after the collapse of activation the cells enter a refractory period. Either the excess of inhibitor must decay or the substrate becomes exhausted and must be

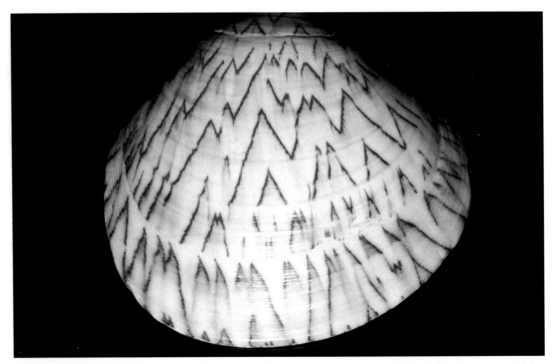

Figure 3.6. Shell pattern generated by travelling waves: pattern on *Lioconcha lorenziana*. The oblique lines result from a process that can be compared with an infection. Pigment-producing cells infect their neighbours such that they also produce pigment, while pigment production is switched off in infecting cells after a short period. The cells become refractory, i.e., they become immune to another infection. Thus pigment production moves along the pigment-producing mantle gland like a wave. In the time record, this leads to oblique lines. A cell that starts spontaneously with pigment production gives rise to two diverging lines (∧-element). If two travelling waves collide, both waves become extinct since a wave cannot enter into the region made refractory by the counter wave. Such mutual annihilation leads to ∨-like pattern elements.

replenished (see Figure 3.3). The inclination of the oblique lines is determined by the ratio of the speed of the waves and the speed of shell growth.

After spontaneous activation in a small group of cells, both neighbouring cells can be infected. At this point two waves are initiated that run in opposite directions. Such an event causes a ∧-like pattern with two diverging oblique lines. Conversely, if two waves collide, they annihilate each other since waves cannot enter into a region made refractory by a counter wave. On shells, this leads to ∨-like pattern elements (Figure 3.6).

For travelling waves, activator diffusion must be within a certain range. If it is too high, an overall synchronization of the oscillations will occur and stripes parallel to the edge will result. On the other hand, if it is too low, oscillations of individual cells will become independent of each other and the phase relation among the cells will be lost. If the antagonist diffuses at all, the diffusion rate must be much lower than that of the activator, otherwise the faster spreading antagonist would arrest the wave.

A pure travelling wave mechanism for shell patterns seems to be more the exception than the rule. The shell in Figure 3.6, for instance, shows many details indicating that the actual mechanism is more complex. Oblique lines terminate without prior collision, and many of these terminations take place simultaneously, indicating some global influence. At other locations one of two lines terminate shortly before wave collision takes place, indicating the long-range influence of one wave on another. While the upper boundary of the oblique lines appears smooth, the lower side can have small dents or a smear. Later we will see that these features are expressed even more strongly on other shells, providing important clues to the understanding of more complex patterns.

3.4   Travelling waves require a pace-maker region

In order for travelling waves to occur, the waves must be initiated at particular positions. Otherwise synchronous oscillations may occur even though the general conditions for travelling waves are satisfied (Figure 3.7a). One possibility to explain the initiation of travelling waves is that oscillation frequency depends on the position. Faster oscillating cells may become pace-maker regions. The pace-maker role of the sinoartrial node in the contraction waves of the heart muscle is a well known example (see Winfree, 1980, Glass and Mackey, 1988). Due to its higher oscillation frequency, a pace-maker region fires earlier than its surroundings, initiating travelling waves at specific positions. It may require many rounds of oscillation before the dominating role of a pace-maker region becomes fully established (Figure 3.7c). As discussed below the formation of a pace-maker region may require a separate pattern forming process.

Pace-maker regions can also appear spontaneously if parameters slowly change the system from the excitable to more sustained oscillations. Those cells that first start to oscillate trigger their still excitable but not yet spontaneously firing neighbours, starting travelling waves. If the cells are on the borderline between the arrest of a cycle in the excited state and sustained oscillations (see Figure 3.2b), relatively minor fluctuations can lead to dramatic differences in oscillation frequencies and thus to pacemaker regions (Figure 3.7b).

Strong support for the suggestion that oblique lines are generated by travelling waves can be derived from pattern regulations frequently seen on the shells of *Strigilla carnea* (Seilacher, 1972, 1973). The normal pattern consists of very regular oblique ridges that merge along a particular zone, producing a pattern of nested V's (Figure 3.8). In some specimens, the normal pattern has been perturbed, probably by an external event. Some neighbouring waves become terminated causing a gap in the pattern of oblique lines. Pattern regulation occurs either by a bending of the remaining lines towards the gap and/or by the spontaneous initiation of new lines. The computer simulation in Figure 3.8 shows that the model correctly describes the essential features of this pattern regulation: the bending takes place only in an

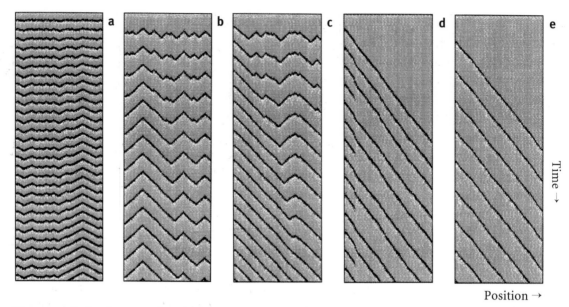

Time →

Position →

Figure 3.7. Role of pace-makers in the generation of travelling waves, calculated using the activator-depletion mechanism in Equation 2.4. (a) If the oscillation frequency is the same in all cells, oscillations can remain synchronous even though the conditions for travelling waves are satisfied. (b) If some cells at random positions oscillate somewhat faster than most other cells, these faster cells form pace-maker regions. These regions are the periodic initiation points of two diverging lines since a spontaneously activated cell can infect both neighbours (∧-like pattern element). (c-e) Very regular oblique lines emerge only if a particular group acts as a pace-maker. (c) The substrate production b_b is 30% higher in the leftmost cell. It may require some time to bring a larger field under the control of a pace-maker region. (d) If a pace-maker fires too rapidly, not all initiated waves survive. This leads to termination of oblique lines. (e) Except for the pace-maker cells, the cells are unable to trigger spontaneously but they can propagate the excitation. Ordered travelling waves are formed from the beginning [GT37; S37A shows the situation in (c) in a movie-like manner].

upward direction resulting from a speeding up of remaining waves; new initiation points lead to a W-like pattern, a pattern which is otherwise absent on the shell.

The model predicts that the initiation zone of the travelling waves (∧'s) may have special properties indicating a pace-maker region. In contrast, the annihilation zones (∨'s) are arbitrary and depend only on where the waves collide. This is supported by another pattern abnormality observed on these shells (Figure 3.9). Due to an interruption of some waves, the zone of mutual annihilation, i.e., the position of the tips of the ∨'s, can be shifted.

In conclusion, oscillations result when self-enhancing processes are antagonized by processes too slow to enable a stable balance. On shells, this can lead to stripes either parallel or oblique to the growing edge.

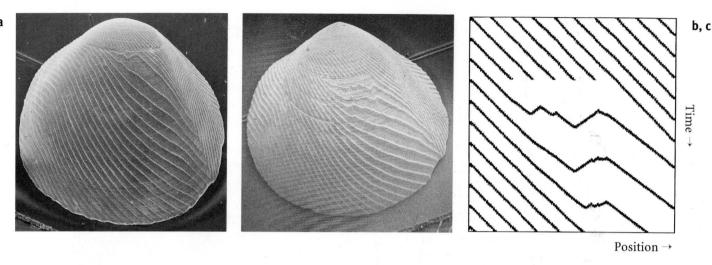

Figure 3.8. Pattern regulation: repair in the rib-like pattern of *Strigilla*. (a,b) Examples: after a perturbation of the normal pattern (caused presumably by an external event) some lines are interrupted but the pattern discontinuity disappears over the course of time. (c) Model: The oblique rib-like pattern is assumed to be generated in a similar way to the corresponding pigmentation pattern, i.e., by travelling waves. Interruption is simulated by an artificial lowering of activator concentration. Due to the absence of activation for a prolonged period, the substrate concentration (not shown) increases to such a level that a spontaneous activation becomes possible (∧ -element). Due to the accumulating substrate, the cells become more and more susceptible to triggers from neighbouring cells. This leads to a speeding up of the travelling waves and the bending of the lines. The bending is unilateral since, of course, the waves on the other side of the gap are not retarded [GT38].

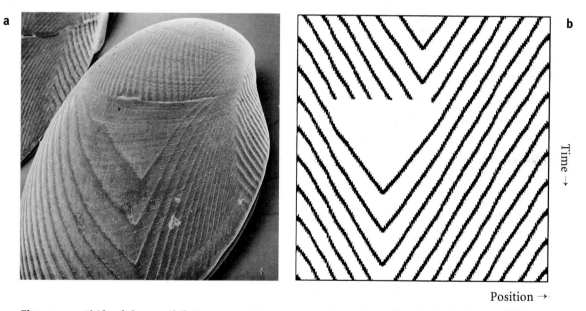

Figure 3.9. Shift of the annihilation zone. (a) A pattern irregularity that includes the annihilation zone (V-elements). (b) Model: If the gap is asymmetrically located with respect to the normal annihilation zone, the new region of annihilation is shifted accordingly. This supports the view that the position of the nested V's is determined only by where the waves collide, not by special properties of the corresponding cells. A bending of the ridges towards the gap can take place on both sides [GT38]; (Specimens kindly provided by A. Seilacher, see Seilacher, 1972, 1973)

Figure 4.1. Superposition of two patterns, a periodic pattern in space that is stable in time and a periodic pattern in time that is spatially homogeneous. At the top, the shell of *Ficus gracilis*. Both patterns coexist without much interference. At the bottom: *Bursa rubeta*. The pattern stable in time determines where, the oscillating pattern determines when the spines will be formed.

Chapter 4

Superposition of stable and periodic patterns

A widely distributed subgroup of shell patterns result from the superposition of a stable and a periodic pattern. The upper shell in Figure 4.1 shows two sets of parallel relief-like lines. One set is oriented parallel to the growing edge and results from a thickening of the shell at periodic time intervals. The other set is oriented parallel to the direction of growth and results from a permanently enhanced deposition of shell material at regularly spaced positions. In this example, the two patterns do not interfere with each other, a situation that is more the exception than the rule, but it shows that the assumption of two superimposed systems is reasonable.

More intriguing are systems in which one system modifies the other. The bottom shell in Figure 4.1 shows spines that have been formed at regular time intervals in defined positions. Evidently, a stable pattern must exist that determines the position of the spines, and an oscillating pattern must decide the time at which spine formation actually occurs. Another regular substructure can also be recognized. Every second row is different from the intervening row. The first spine, counted from the shoulder, appears only in every second row and is much larger. After every second row a discontinuity in the shell is formed. Thus, with its combinatorial possibilities, superpositions of two (or more) patterns can provide a rich source of complexity.

This chapter will discuss patterns that result from the modulation of parameters in an oscillating system by a stationary pattern. Usually the modulating pattern is not directly visible but has to be inferred from the space-dependent behaviour of the oscillating system. The diversity in patterns results from differences in the actual form of the stable pattern, from its action on the oscillatory system, as well as from the particular properties of the oscillating system. The following cases will be considered:

(i)   The oscillation frequency is space-dependent resulting in a pattern of wavy lines.

(ii) Steady state activations occur in some regions while in other regions the system oscillates. Fishbone-like patterns result.

(iii) Oscillations are restricted to particular regions, while in other regions they do not occur. The resulting pattern exhibits rows of dots or crescents.

4.1   The formation of undulating lines and the partial synchronization of cells by activator diffusion

A broad spectrum of patterns from very different species may be generated by oscillatory pigment deposition in which the frequency of the oscillations varies along the growing edge in a systematic manner. This feature is obvious on the shell of *Natica euzona* (Figure 4.2). For speed and simplicity in the following simulations, the stable pattern is not calculated but supposed to exist. Usually, it is shown on top of the simulation. The generation of stable patterns was discussed earlier in some detail (see Figure 2.3 and Figure 2.5). Its explicit integration into the models will be exemplified later in this chapter (Figure 4.13).

In the activator-substrate model higher substrate production rates lead to higher oscillation frequencies (until the system enters into a steady state of activation). Thus, if a stable pattern exists that modulates the substrate production b_b (see Equation 2.4, page 28), the oscillation frequency becomes space-dependent. Initially, if all cells are in the same phase, subsequent pigment production will occur earlier in the faster oscillating cells due to the modulating pattern. Thus, the next line will have a wavy appearance. In the absence of any activator diffusion, the phase difference between neighbouring cells will increase with each further oscillation until any phase relation is lost. Diffusion of the activator has the tendency to synchronize neighbouring cells since cells activated earlier advance their delayed neighbours by activator exchange. Thus, travelling waves are possible. Cells in regions of high substrate supply act as pace-makers due to their faster oscillations.

Travelling waves can proceed only if the phase difference between neighbouring cells is not too large. The cell to be triggered must be in a sensitive phase in order for the addition of small amounts of activator to release the autocatalytic burst. With the activator-substrate model, this means that substrate concentration must have reached a certain level. Otherwise the chain of triggering events will be interrupted and a pigment line on the shell will terminate. Figure 4.3 shows this process in detail. For example, the first wave will become slower as it enters a region of lower substrate production but it may find sufficient substrate to survive. However, the situation becomes more critical for subsequent waves. Since the speed of the fore-running wave is reduced, less time is available for the recovery of substrate concentration, especially since its production is reduced at this point. The available substrate may be insufficient and wave termination will occur. Thus, wave termination is expected to occur in regions where neighbouring cells have different oscillation frequencies. The faster oscillating cells can no longer entrain the slower oscillating cells. In the model, a slowing down of the wave is expected before wave termination occurs. On the final shell this causes a bending of the pigmentation line towards the growing edge.

Connected with the termination of a pigmentation line is an interesting pattern irregularity on the subsequent pigmentation line. This line frequently shows a small

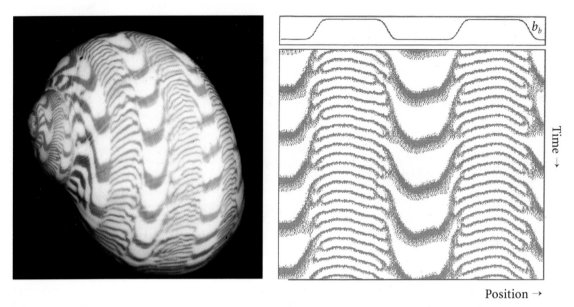

Figure 4.2. Shell of *Natica euzona*. The oscillation frequency is much higher in some regions, indicating the existence of a spatially stable pattern that influences the oscillation frequency. Model: space dependent substrate production $b_b(x)$ (see Equation 2.4) leads to different oscillation frequencies. Lines of pigmentation are formed with different spacing. In the transition zone lines can merge or end blindly. To model the narrower bands in the region of high oscillation frequencies, the decay rate of the activator r_a and the source density s are assumed to have the same spatial pattern as the substrate production rate b_b [S42].

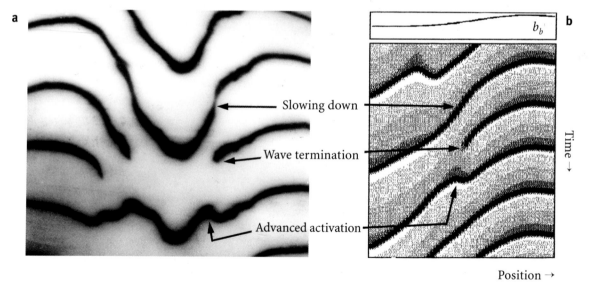

Figure 4.3. Termination of wavy pigmentation lines. (a) Example from *Amoria macandrewi*. (b) Model: after a travelling wave enters a region of low substrate production, the substrate concentration (red) may be too low for the wave to survive. The subsequent pigmentation line shows a characteristic irregularity [GT44]

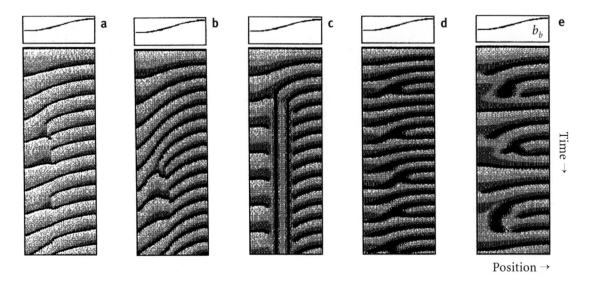

Position →

Time →

Figure 4.4. Modulation of an oscillating activator-substrate system by space-dependent substrate production (curve at the top), and the influence of diffusion and saturation. A region of higher substrate production acts as a pacemaker. (a) Low activator diffusion leads to waves that travel into a region of lower substrate production. The waves may terminate (see also Figure 4.3). (b) At higher saturation ($s_a > 0$) the activation period is longer (thicker lines). More time is available to infect neighbouring cells. Slower waves (steeper lines) are possible. (c) Diffusion of the substrate can lead to the coexistence of stable and periodic activations. Stable activations occur preferentially in a region where neighbouring cells show pronounced differences in the oscillation frequency. (d) Higher diffusion rates of the activator and the substrate can cause bifurcating pigmentation lines. (e) A lower substrate production causes more terminating lines and incomplete forks [GT44].

Figure 4.5. Wavy lines. (a) The shell of *Amoria undulata* (top) and *Amoria marcandrewi* (bottom). (b, c) Simulations: the faster oscillating cells entrain the slower oscillating cells. If the phase difference is not too large, a newly activated cell can activate its somewhat delayed neighbours. The steepness of the lines is an indication of the readiness of a cell to become activated. If the phase difference becomes too large, triggering neighbouring cells is not possible and the pigmentation line will terminate abruptly. Neighbouring cells skip one oscillation. After a long period without triggering, an advanced spontaneous activation may occur, in agreement with the natural pattern. (c) With a higher saturation, the tendency for line termination is reduced and long, steep oblique lines result [S45b, S45c].

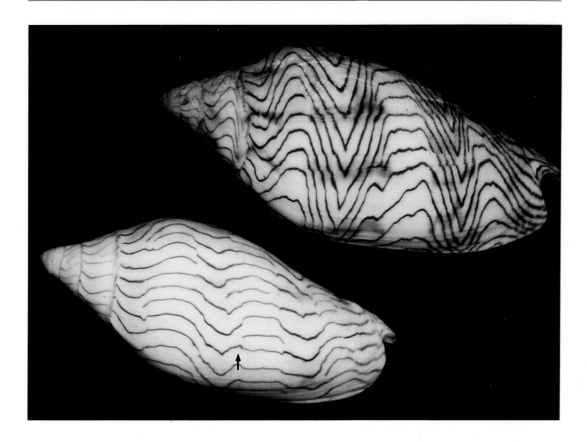

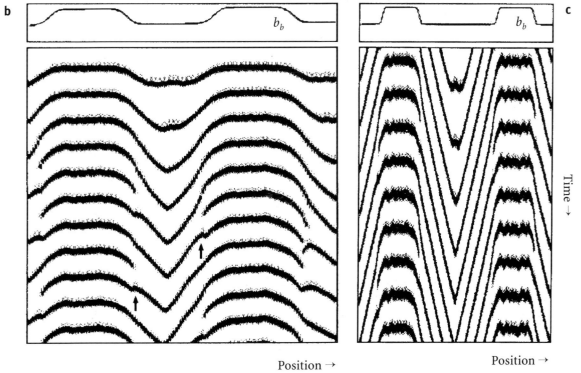

b

b_b

c

b_b

Position →

Position →

Time →

∧-like bending towards the terminated line. The effect is reproduced in the model and may be explained as follows. In front of the terminated wave the substrate will not be used up. This locally higher substrate concentration leads to an earlier activation. The situation is illustrated in detail in Figure 4.3. Figure 4.5 shows a simulation on a larger field as a comparison with the corresponding shell.

4.2   Reducing wave termination with a longer activation period

Wave termination also depends on the relative length of the activated period. If activation persists for a relatively long fraction of the oscillatory cycle, the chances are good that enough activator will be transmitted to the neighbouring cell by diffusion to activate it. If the increase in activator concentration is limited due to saturation, then the maximum substrate consumption is also limited. The same "fuel" is sufficient for a longer time period and the width of the pigmentation lines increases (see Figure 3.3d). As shown in the simulation in Figure 4.4b, saturation largely prevents line termination. Or, conversely, if the activation period is long, the exchange by diffusion can be much smaller and still maintain the chain of triggering reactions. In such cases, the pigment lines can be very steep. An example of this situation is given in Figure 4.6a, b. Figure 4.6c, d illustrates schematically the connection between line width and maximum steepness.

4.3   Interconnecting wavy lines and the formation of arches

In some species connections between subsequent lines do occur, either regularly (Figure 4.2) or occasionally (Figure 4.7, 4.8). As a rule, these connecting lines are steep and thin. The steepness indicates that the speed of the travelling waves in such regions is very slow. According to the model, these interconnections result from a moderate diffusion of the substrate (the antagonist). Low level activation can be maintained due to the influx of substrate from the surrounding cells when otherwise wave termination would occur. However, since some substrate has already been consumed by the cells into which the wave must spread, the activator concentration is low (and the pigmentation faint). More time is required before a neighbouring cell is triggered, i.e., the speed of the wave is reduced and the line is steep. During this slow movement of the activated zone, the substrate concentration increases in neighbouring cells. Later, full activation will be triggered again in those cells in which activation survived at low levels. On the final shell weak but steep lines are formed that connect two subsequent pigmentation lines (Figure 4.7).

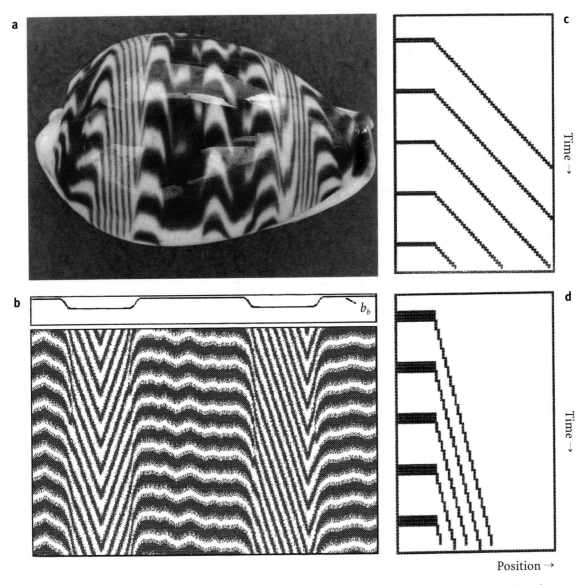

Figure 4.6. Steep lines. (a) Shell of *Cypraea diluculum*. (b) Model: if the activator autocatalysis has an upper bound (saturation), the activated period within a cycle is relatively long and the stripes are thick. If activator diffusion is low, the activation of one cell by its activated neighbour requires time. Since the activated portion of the cycle is long, more time is available in which one cell can infect its neighbour. Despite the large phase difference, this does not lead to wave termination. A large phase difference can accumulate between neighbouring cells. The result is thick lines in regions with high oscillation frequencies and very steep but narrow lines in regions with lower oscillation frequencies. (c,d) Schematic drawing to illustrate the connection between line width and maximum steepness; [S46].

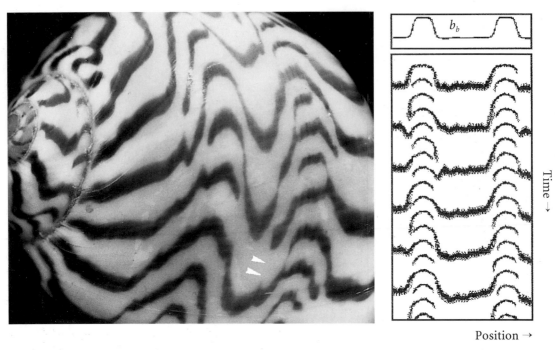

Position →

Figure 4.7. Connected arches: shell of *Natica undulata*. Model: if, in addition to activator diffusion, some substrate diffusion takes place, the connection of one line with the following line is possible. Due to substrate supply from the surrounding cells, the activation can be maintained almost locally. The travelling waves can come close to rest without becoming extinct. This "surviving" activation triggers activation of the surrounding cells as soon as their refractory period is over. The resulting pattern consists of a steep line forming a connection between subsequent pigmentation lines. The two arrowheads mark a hidden line [S47].

4.4 Hidden waves

On shells with arches, line termination may also occur occasionally. But the mode of re-activation indicates that the mechanism is somewhat different from the line termination mentioned above. The re-activation occurs at the elongation of a terminated line, i.e., at a displaced position (see the two arrows heads in Figure 4.7). This indicates that the travelling wave was still present, although the activator concentration was too low to evoke pigmentation. This observation suggests that pigment production is not by itself the autocatalytic process, since wave propagation can continue without (visible) pigmentation. Wave formation may be an independent process. Only if the signal is high enough would pigmentation be accomplished. Alternatively, it may be that wave propagation proceeds at two different levels (see Fig. 8.8) and the lower level is insufficient to evoke visible pigmentation.

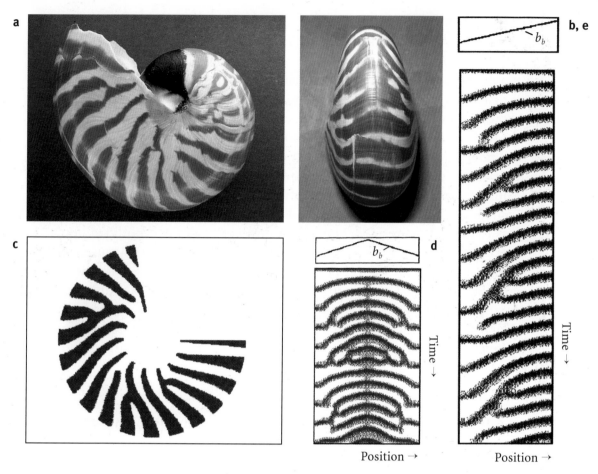

Figure 4.8. (a,b) Shell of *Nautilus pompilius* in two views. The number of lines at the shell periphery is approximately double the number in the inner region, indicating a gradient in the oscillation frequency. Model: an activator-substrate model with a (linearly) graded substrate production $b_b(x)$ is assumed. The same gradient may also be used to maintain the higher growth rate on the periphery which causes the spiral shape of the shell. (c) Pattern drawn as partial circles to demonstrate the similarities between the natural pattern and the simulation. (d) Simulation of the peripheral view. In the region of high substrate production, activator production attains a nearly steady state. (e) Calculation as in (c) in normal space-time plot [S48c, S48d].

4.5 Pattern on the shell of *Nautilus pompilius*

The spiral shape of the *Nautilus* shell results from a much faster growth along the periphery of the growing edge. Ward and Chamberlain (1983) have determined the growth rate of animals kept in the aquarium to be 0.15 to 0.25mm per day (or 7cm per year). Accordingly, the animal shown in Figure 4.8 produced about 8 pigmentation lines on the peripheral region of its shell in the last year of its life. At the same time only 4 lines have been formed on the inner region. This indicates roughly a factor of two in the oscillation frequency between the inner and outer side of the shell. Irregularities on the surface of the shell provide a record of the shape of the growing edge over the course of the animal's individual history.

These growth lines result neither from a daily nor a lunar rhythm. Saunders (1984) found a substantial variation of 4 to 15 days per growth line in the larger species *Pompilius belanensis*.

A stable pattern must exist that controls oscillation frequency. It must have a graded distribution. For the sake of simplicity in the simulation in Figure 4.8c, a linear gradient has been assumed. To obtain the spiral shape of the shell, a similar gradient in the growth rate is required. Both oscillation frequency and growth rate may be under the control of the same gradient.

The pattern on the *Nautilus* shell shows pattern elements similar to those discussed in the previous section: interconnection of successive pigmentation lines, and terminating lines which are restricted to a region of high oscillation frequency. It is thus tempting to assume a similar mechanism to the one outlined above. In this case, moderate activator diffusion leads to partial synchronization among neighbouring cells. High saturation causes the pigmentation lines to have about the same width as the space in between. Due to moderate substrate diffusion, two pigmentation lines can fuse. Termination of pigmentation lines also occurs. The similarity between the simulated and the natural pattern becomes especially striking if activator production is plotted as partial concentric circles corresponding to the actual mode of shell growth (Figure 4.8).

4.6 Stabilizing an otherwise oscillating pattern by diffusion

As outlined above, oscillations occur if either substrate production is too low to maintain a permanent activation or the inhibitor reaches equilibrium too slowly due to low turnover. However, the situation may change if the antagonist strongly diffuses, even though all time constants remain the same. An additional substrate supply from the surroundings via diffusion can be sufficient to maintain a steady state activation. In the same way, a substantial loss of inhibitor by diffusion has an effect similar to shortening the inhibitor's life and can equally cause the transition to a steady state. In such cases, lateral inhibition comes into play. The transition to the steady state can occur only with a certain spacing. Figure 4.4 and Figure 4.7 illustrate the formation of connections between subsequent pigmentation lines based on substrate diffusion. Figure 4.9 shows that even stronger substrate diffusion can cause a localized transition into steady state activation. On the shell this leads to lines parallel to the direction of growth. Between these lines the oscillations can continue. Line formation is expected to occur preferentially in regions with larger differences in oscillation frequency since the onset of stabilization requires a phase difference between neighbouring cells. In this case an activated cell can profit from its non-activated neighbour. The snail *Voluta musica* in Figure 4.9 produces a shell with a related pattern. Fine lines perpendicular to the growing edge are interspersed with periodic patterns of dots and fine lines parallel to the growing edge. These pattern elements are reproduced in the simulation. The sim-

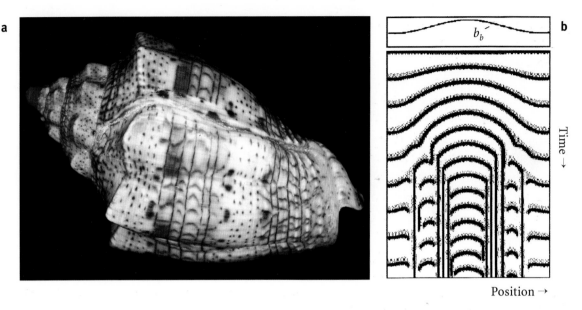

Figure 4.9. (a) Lines parallel and perpendicular on the same shell - the shell of *Voluta musica*. The pattern is characterized by two bundles of fine lines perpendicular to the growing edge, interspersed by periodic pattern elements. (b) Simulation. In regions of high substrate production (high b_b in Equation 2.4), the system is close to the transition from oscillating to steady state activation. Diffusion provides an additional supply of substrate from neighbouring non-activated regions. In this way steady state activation emerges at several sharply localized positions [S49].

ulation is certainly not perfect since fine dots are formed on the shell between the bundles of lines while in the simulation more lines appear. The simulation merely shows the diffusion-based transition from an oscillatory regime to localized steady state activations.

4.7 Combinations of oscillating and non-oscillating patterns

Very different patterns emerge if the stable pattern shifts groups of cells into a permanent steady state. This occurs, for instance, if the local substrate production surpasses a certain level. Permanent pigment production by these cells leads to stripes parallel to the direction of growth. The cells in between these stripes are still able to oscillate. Travelling waves are initiated at regular distances by the permanently activated cells. The resulting pattern consists of nested V's between the stripes. Figure 4.10 shows a simulation together with the shell of *Cypraea ziczac* for comparison. The peripheral pattern of a *Nautilus* shell (Figure 4.8b,d) was an example of a system on the borderline between oscillation and steady state.

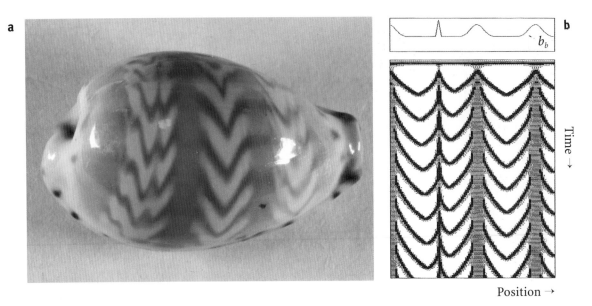

Figure 4.10. Fish-bone pattern resulting from a combination of permanent and oscillatory pigment deposition. (a) Shell of *Cypraea ziczac*. (b) Model: an activator-substrate model is assumed. In regions where substrate production is high, cells enter into steady state activation, causing pigmented bands parallel to the direction of growth. These permanently activated cells periodically initiate waves travelling in the space between the bands. The V-like elements result from the annihilation of pairs of waves at the point of collision [S410].

4.8 Rows of patches parallel to the direction of growth

An even more common pattern is the reciprocal of that discussed above. Oscillations take place only in particular regions on the shells. Rows of patches, dots, short lines or crescents result. They are separated by bands in which no pigmentation occurs.

It is tempting to assume that this pattern may be generated by an elementary system in which the antagonist has a high diffusion rate (pattern in space) and a large time constant (pattern in time). However, as mentioned, the mechanism for lateral inhibition requires some time before a particular region will dominate over a neighbouring one. In oscillating systems with burst-like activation this time is not available (see Figure 3.6). Again, a stable pattern must be responsible for determining where activations will take place.

It is possible that a periodic pattern will generate its own stable pattern due to positive feedback from the activation on the source density, i.e., on the general ability of a cell to perform autocatalysis (s in Equation 2.1 and 2.4). Despite the initially sloppy separation of the patches due to a long-lasting influence on the source density, regions of slightly enhanced activation will obtain a stronger advantage in the next oscillation and so on until a stable periodic pattern in the source distribution emerges. A simulation is provided in Figure 4.11. Feedback by an activator pattern on the long lasting source density seems to be a general

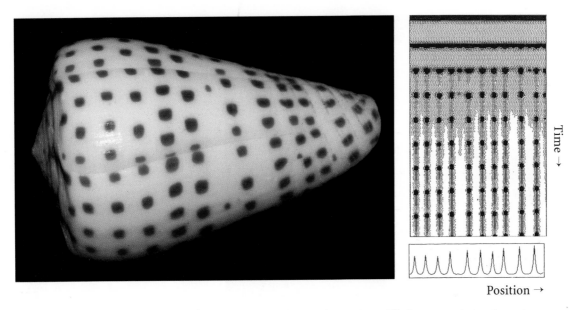

Time →

Position →

Figure 4.11. Generation of a stable pattern by feedback from an oscillating pattern on the source density. In an oscillating system rapid diffusion of the antagonist does not lead to well-resolved patches. With the feedback of activator concentration on the source density (green), a slightly stronger activation leads to an even stronger activation in the next cycle and so on until rows of patches emerge along the stable periodic pattern of the source density (curve at the bottom). However, as shown in Figure 4.13, the actual mechanism presumably involves two complete systems; [S411].

mechanism in development and is used, for instance, for the generation of stable tissue polarity. If a monotonic activator distribution is generated (see Figure 2.9) and the activator produces feedback on the source density, the source density will obtain a graded distribution too. In such a system, once the activated region is removed, a new activator maximum regenerates in the remaining tissue at the point of highest source density, i.e., at the side pointing towards the original activated region. Such a "memory" of the tissue is suggested by classical experiments with the fresh water polyp *Hydra*. Pieces of *Hydra* tissue always regenerate a new head on the side that points towards the original head (see Gierer, 1977, Meinhardt, 1993).

The types of patterns that can be generated by this mechanism are limited. For example, rows consisting of large patches in close proximity cannot be produced since the required saturation together with the required inhibitor diffusion would lead to continuous stripes instead of isolated patches. These restrictions disappear if, as in the examples discussed above, two patterns are superimposed. Two modes are conceivable for the action of the stable pattern. Either the stable pattern generates the precondition that allows oscillations to take place or, conversely, the conditions for oscillation are given everywhere except in those regions in which the stable pattern is in an activated state. One possible way to distinguish between these two modes is by the spacing between the rows. The activated region is usually smaller than the non-activated region. On the shell shown in Figure 4.12 only

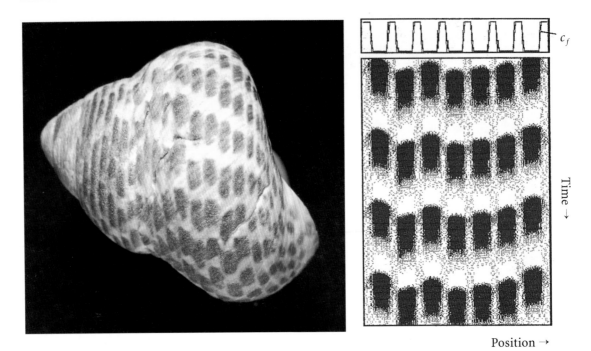

Figure 4.12. Narrow permanently unpigmented lines are indicators of an active suppression of the pigmentation reaction. Shell of *Austrococlea adelaidae*. Activated regions are usually much smaller than the non-activated regions in between (see Figure 2.3). Therefore, the stable pattern must cause the suppression of pigmentation reaction and cannot function as a precondition for pigmentation. Simulation: constant pigmentation over a long period of time can be achieved if a separate extinguishing reaction is involved (shown in red) that terminates pigment production (see Figure 7.5); [S412]

very narrow pigment-free lines separate the much broader regions in which oscillatory pigment production actually occurs. The fact that the activated regions are usually smaller than the non-activated ones, suggests that a localized suppression of oscillation by a spatially stable pattern is involved.

Figure 4.13 shows a simulation that assumes two activator-inhibitor systems. The stable system (the activator shown in red) acts as an additional inhibitor on the oscillating system (black). Thus, black patches can appear over the course of time only along the (non-red) interstices allowed by the stable system. Partial synchronization can take place such that several patches are formed at the same time.

On some shells the distances between the rows of pigmentation are irregular. A possible cause of these irregularities is that the interstices must have a certain width in order for periodic activations to occur. If two maxima of the stable pattern are too close together, the concentration in between is too high to allow the onset of oscillations. However, with the growth of an animal, the size of the interstices increases too and oscillations become possible. On a shell, such an event leads to the insertion of a new row of patches in a larger, formerly non-pigmented region. On the other hand, from a certain distance between the maxima of the suppressing pattern onwards, the insertion of new maxima is expected (see Figure 2.6 and 2.7).

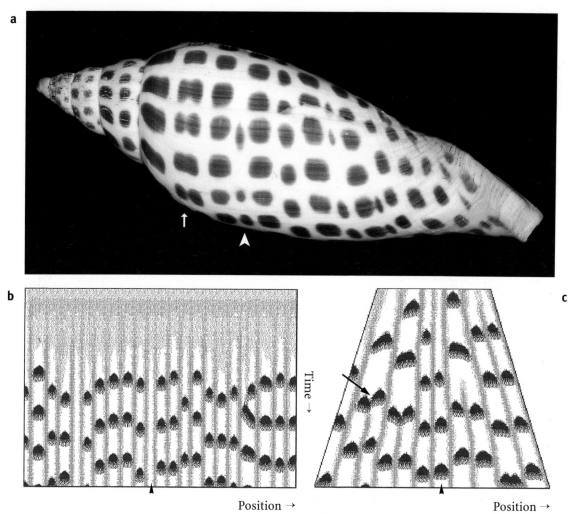

Figure 4.13. Rows of patches. (a) Shell of *Voluta junonia* (b) Model: A stable (red) and an oscillating (black) activator-inhibitor system are assumed. By cross-inhibition, the stable activator suppresses ctivation of the oscillating system. Oscillations may only occur if the space between the suppressing stable pattern has reached a certain minimum size. The arrow marks an empty space. (c) Simulation of a growing system. After sufficient growth a new row of dots is inserted (arrowheads in the simulation and on the shell). After surpassing a critical size, the suppressing system may insert a new maximum (arrows), causing the split of a row into two [S413B, S413C].

On the shell, a row with large patches of pigmentation will split into two smaller rows. Both the split and the insertion of pigmentation rows can be seen on the pattern of *Voluta junonia* and are reproduced in the simulation (Figure 4.13). Such patterns demonstrate that the stable pattern is not fixed during an early stage of development, but that it is dynamically regulated in the growing animal.

4.9   The possible role of a central oscillator

Generation of the three-dimensional structure requires different growth rates along the mantle gland of the shell (see chapter 10). A very remarkable feature in many shell patterns is the constancy of the spacing between periodic pattern elements despite the different speed of accretion of new material at the shell's growing edge. One possible mechanism has been discussed already for the *Nautilus* shell (Figure 4.8): the growth rate and the oscillation frequency may be under the control of the same graded signal.

The constancy of spacing implies that more lines are inserted at positions with higher shell circumference. When, for instance, in a region of larger shell circumference ten, and in a smaller region only nine stripes are inserted in a certain time interval, there must be a zone of mismatch. An inspection of the shells tells us that these phase conflicts between neighbouring oscillators are resolved in a very short time interval, within one or two oscillations. On the shells this causes, for instance, one line to end blindly while the preceding and subsequent lines make connections (Figure 3.1 and 3.4). Remaining phase differences are resolved by a fast re-synchronization of the oscillators, as indicated by a sharp bending of the corresponding lines. Similarly, in Figure 4.14, a single line makes a sharp bend and fuses with the subsequent line.

Attempts to simulate this phenomenon turned out to be more difficult than expected. When the differences in the oscillation frequencies are small but the tendency to mutual synchronization is large (as indicated by straight parallel lines), it is expected that synchrony is maintained, as in the case of the shell shown in Figure 1.4. Line termination as demonstrated in Figure 4.4 is difficult to obtain under these conditions. If a diffusing inhibitor is involved in the synchronization, line termination can occur due to lateral inhibition (Figure 4.14b). However, the resulting pattern differs substantially from the natural pattern. It requires several oscillations in the zone of conflict before synchronization is re-established.

An alternative possibility is shown in Figure 4.14c. In addition to the oscillating pigment system (black), an endogenous oscillating system is assumed (green) that is homogeneously distributed in the organism in a hormone-like manner. It oscillates with approximately the same frequency as the pigment system and has an enhancing influence on it. Therefore, the pigmentation system will be entrained by the central oscillator as long as the phase difference is not too large. Otherwise the pigmentation system will skip one pulse of the central clock and becomes entrained to the next pulse. As Figure 4.14c demonstrates, the simulation reproduces space-dependent reduction of line density much better. Re-synchronization is accomplished within one or two pulses. The synchronization remains through many more pulses until the next short relief of phase differences occurs.

Evidence for an oscillating system independent of the pigmentation system but with a stimulating influence on it was provided earlier in Figure 1.10, where the periodic modulation of the background pigmentation and the locally restricted periodic pigment production can be easily distinguished. Both systems are in partial synchrony. With the analysis of triangular pattern elements in chapter 8, another example of the influence of global oscillations on pigment production will be discussed.

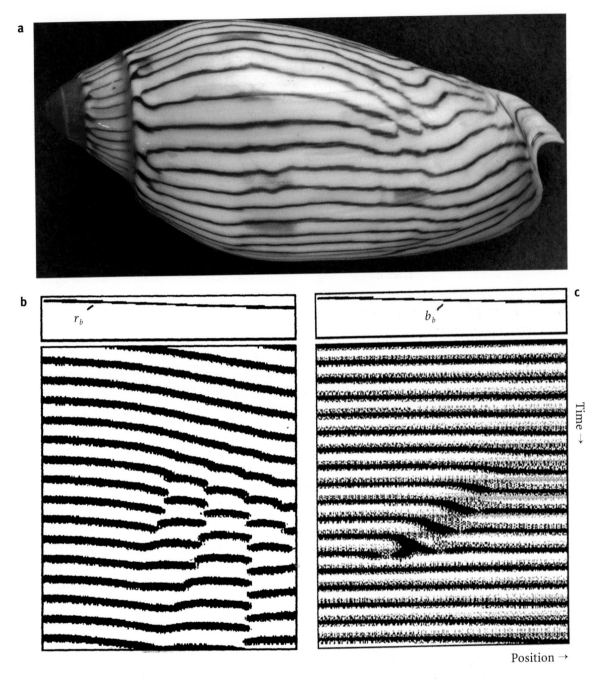

Figure 4.14. Synchronization of oscillating pigment deposition by an internal clock. (a) Shell of *Amoria turneri*. The lines have a constant spacing. The number of lines is reduced discontinuously in regions of smaller shell diameter. (b) With a diffusing inhibitor and a space-dependent oscillation frequency (lifetime r_b is space-dependent), accumulating phase differences between neighbouring oscillators can cause periods of re-synchronization between adjacent cells. This would lead to extended zones of line mismatch. (c) A central oscillator (green) has a stimulating influence on the oscillating pigment system. Whenever the phase difference becomes too large, a local resetting is possible within one or two pulses, which is in much better agreement with the natural pattern [S414b, S414c].

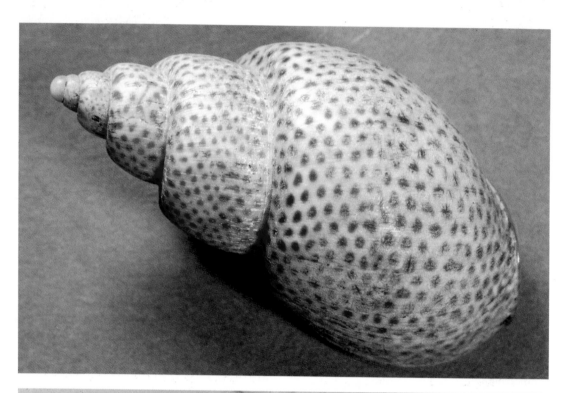

Figure 5.1. Staggered dots on *Babylonia papillaris* and oblique lines with crossings on *Tapes literatus*. Although the patterns look very different, it is proposed that both are based on a common mechanism. Two antagonists enforce a periodicity in space and time.

Chapter 5

Crossings, meshwork of oblique lines and staggered dots: the combined action of two antagonists

Many shells display simple periodic patterns that cannot be accounted for with the elementary mechanisms described so far. Patterns of staggered dots and mesh-works belong in this class (Figure 5.1). These patterns are characterized by a periodicity along the time coordinate as well as along the space coordinate. This suggests that two antagonists are involved: a non-diffusible one that is responsible for the periodicity in time, and a second highly diffusible one that causes the pattern through space. The interactions described by the equations in this chapter are possible extensions of the activator-substrate and the activator-inhibitor model (see boxes). An important property of such mechanisms is that travelling waves can penetrate each other without annihilation. In other words, crossings of oblique lines can occur.

5.1 Displacement of stable maxima or enforced de-synchronization by a second antagonist

For a more intuitive understanding of the role of the second antagonist let us first regard the elementary pattern on its own and then take the action of the second antagonist into consideration. If the primary system would lead to a stable pattern (highly diffusible antagonist), the local accumulation of the second antagonist would destabilize the maxima over the course of time. A neighbouring region, not subject to the non-diffusible antagonistic effect, would become activated. Travelling waves would emerge.

Conversely, a system that is able to generate travelling waves (non-diffusible antagonist) may oscillate in a synchronous manner as long as no pace-maker region is available (see Figure 3.7a). However, a second long-range antagonist enforces a de-synchronization. If a group of cells becomes activated somewhat later than its neighbours, this phase difference will increase during subsequent oscillations. The inhibitory influence that spreads from the advanced neighbours delays the somewhat retarded cells even more. The synchronism breaks down and travelling waves are formed. This type of transition is clearly visible in Figure 5.2.

Equation 5.1 and 5.2: Extensions of the activator-depletion mechanism

For pattern formation involving two antagonists a third substance $c(x)$ is assumed that is produced at a rate proportional to the local activator concentration.

$$\frac{\partial c}{\partial t} = r_c (a - c) + D_c \frac{\partial^2 c}{\partial x^2}$$

It has the following inhibiting effect on activator-substrate interaction (Equation 2.4):

$$\frac{\partial a}{\partial t} = \frac{s\, b}{s_b + s_c\, c}\, a^{*2} - r_a a + D_a \frac{\partial^2 a}{\partial x^2} \qquad (5.1.a)$$

$$\text{with} \quad a^{*2} = \frac{a^2 + b_a}{1 + s_a a^2}$$

$$\frac{\partial b}{\partial t} = b_b - \frac{s\, b}{s_b + s_c\, c}\, a^{*2} - r_b b + D_b \frac{\partial^2 b}{\partial x^2} \qquad (5.1.b)$$

s_b a Michaelis-Menten-type constant. If non-zero, c plays a role only at high c concentrations, while at low c concentration the system behaves the same as a standard activator-substrate system.

s_c efficiency of the additional inhibitor

a^{*2} subsumes activator production by autocatalysis, the baseline activator production b_a and its limitation by saturation

With this interaction, activator concentration (but not substrate concentration) is to a large extend independent of c, since a reduction of c also leads to a compensating decrease in substrate removal. Alternatively, the second antagonistic reaction may result from an additional substrate c that can maintain activator production:

$$\frac{\partial a}{\partial t} = s\, a^{*2} (b + c) - r_a a + D_a \frac{\partial^2 a}{\partial x^2} \qquad (5.2.a)$$

$$\frac{\partial b}{\partial t} = b_b - s\, b\, a^{*2} - r_b b + D_b \frac{\partial^2 b}{\partial x^2} \qquad (5.2.b)$$

$$\frac{\partial c}{\partial t} = b_c - s\, c\, a^{*2} - r_c c + D_c \frac{\partial^2 c}{\partial x^2} \qquad (5.2.c)$$

In both cases, one of the antagonists (b or c) must spread rapidly, while the other should have a low diffusion rate (if any).

a
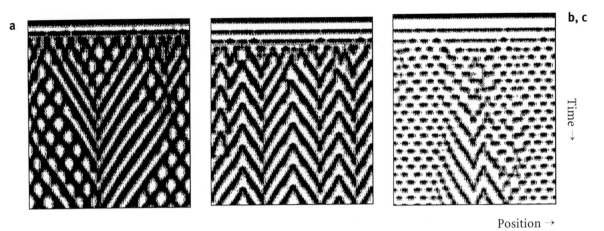
b, c

Time →

Position →

Figure 5.2. Meshwork, oblique lines and rows of dots. These patterns result from the autocatalytic reaction produced by two antagonists, one generating the periodicity in space, the other along the time axis. They are calculated using Equation 5.1 with decreasing rates of substrate production ($b_b = 0.08$, 0.06 and 0.03). In this case, the substrate is the diffusible antagonist. High substrate production leads to permanent activator production whenever the additional inhibitor is low enough and therefore, a meshwork is possible. At lower rates of substrate production only oscillations and thus travelling waves are possible. This results in oblique lines on the shell.[S52].

As shown earlier (Figure 2.7) when two maxima are too far apart, new maxima emerge. In an activator-inhibitor system, new maxima become preferentially inserted some distance from existing maxima. In contrast, in an activator-substrate model, new maxima result from a split and shift of existing maxima. The same features are maintained if two antagonists are involved. As explained below in more detail, the combined action of two inhibitors preferentially leads to isolated patches in a staggered arrangement (Figures 5.11 and 5.12), while the involvement of a depleted substrate leads to shifting maxima and thus, in the time record, to oblique lines and meshworks.

5.2   Pattern variability

The characteristic patterns generated by two antagonists - meshworks, oblique lines and staggered dots - depend on the very different diffusion rates of the two substances. If, in contrast, both antagonists have similar diffusion rates, the overall patterns would be similar to the elementary patterns discussed in chapters 2 and 3, consisting of lines parallel, perpendicular or oblique to the axis. Therefore, the transition from an elementary pattern to a more complex pattern in related species does not necessarily imply the involvement of a different mechanism or an additional substance. Simply, it may be based on a change in the diffusion rate.

The resulting patterns are expected to show a high degree of variability from specimen to specimen. The variation of a single parameter can decide which of these patterns - meshworks, oblique lines and dots - will be formed (Figure 5.2).

Since the ranges of the parameters that form these different patterns will overlap, slightly different initial conditions or random fluctuations can lead to one or the other pattern.

Figure 5.3 shows two shells decorated with oblique lines and a meshwork. These shells were found in different layers of sediment in an ancient lake (Willmann, 1983). They belong to the same family. Although the patterns look so different, it is easy to understand why both patterns result from the same interaction. In both cases a regular alternation between pigmented and non-pigmented regions takes place along the time and space coordinates. In the model, a moderate change in the half life of the diffusible inhibitor is sufficient to accomplish this transition. Therefore, the model provides a rationale for assuming that either a change in environmental conditions or genetic drift has led to the transition from one pattern to the other.

If the travelling waves are organized by pace-maker regions, the resulting pattern is very reproducible since wave initiation takes place at predictable positions (Figure 3.7). In contrast, if the travelling waves are enforced by a second antagonist their starting points and their direction of propagation is not fixed. These ambiguities are revealed in the diversity of patterns seen on different specimens of the same species. Of course, although not necessary, a pace-maker region may be involved in a reaction with two antagonists as well, causing more predictable patterns.

5.3 Global pattern rearrangements

On many specimens a global pattern perturbation can be seen that took place at a particular time. Such an event is indicated by an abrupt and simultaneous termination of a regular pattern that was prominent over a long period of time. A more or less complete rearrangement follows. Figure 5.4 shows two examples in which broad oblique lines disappear in favour of a series of zigzag or wavy lines. Such a global pattern rearrangement is easy to understand by the model. As can be seen in the simulation in Figure 5.3 it takes some time before the oblique line patterns become dominant. An unspecific perturbation, such as a general decrease in activation due to starvation or dryness, can wipe out such an established dominance. For instance, each region in which some activation survives can give rise to two diverging lines, temporarily causing a pattern of wavy lines (Figure 5.4). Somewhat later, lines with a particular orientation can emerge again and the dominating pattern of oblique lines may become re-established. However, the re-established pattern of oblique lines may have a different orientation. Figure 5.6f provides an example of such a situation.

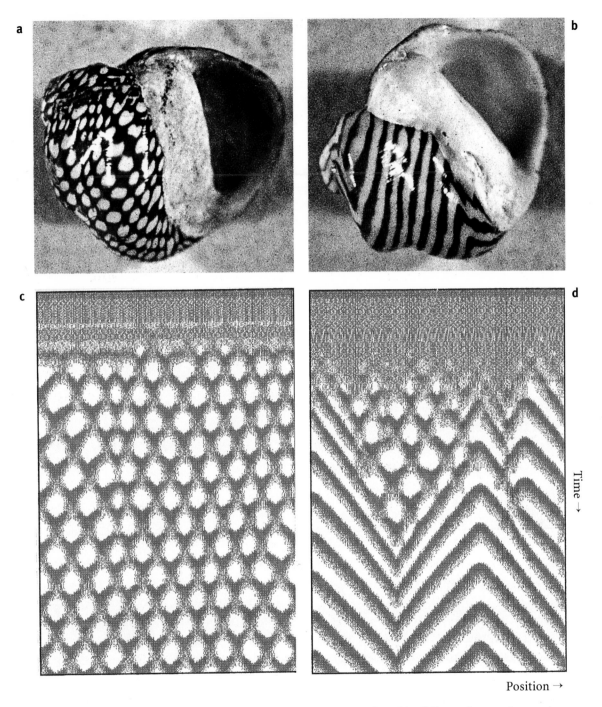

Figure 5.3. Variability of patterns. Shells of *Theodoxus doricus* found in different layers of an ancient lake on the Greek island Koos. Model: the meshwork and oblique line patterns can be generated by the same mechanism. A non-diffusible substrate is responsible for the periodic pattern in time, a diffusible inhibitor for the periodic pattern in space. A shorter lifetime of the inhibitor leads to a more rapid alternation in space. The inhibitory action of one wave onto a subsequent one becomes stronger and the meshwork degenerates into obliques lines. (Photographs kindly supplied by R. Willmann, see Willmann, 1983); [S53].

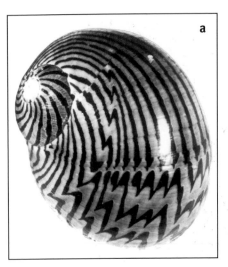
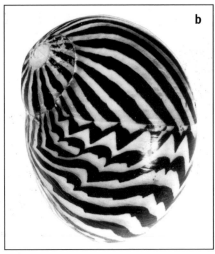

Time →

Position →

Figure 5.4. Pattern transition after global perturbation. (a, b) Two shells of *Neritodryas dubia* GMELIN with an abrupt global transition from oblique lines to a zigzag pattern. (c) In the simulation, a global reduction of the activator leads to the initiation of many new divergent lines (www-pattern). After a while, the oblique line pattern would be restored; [S54]. (Photographs kindly supplied by G. Oster)

5.4   Traces of the additional inhibition: oblique lines initiated or terminated out of phase

A species with an especially rich pattern variation on its shell is the small snail *Bankivia fasciata* (Figures 5.5 and 5.6, see also Ermentrout et al, 1986). These shells show peculiarities in their patterns that very conveniently illustrate the action of a second long-range antagonist.

If travelling waves are based on the elementary mechanism described earlier, no irritation is visible before two waves annihilate each other in a collision (see Figures 3.6-3.9). In contrast, travelling waves enforced by an additional inhibition can behave differently. Shortly before a collision takes place each wave enters a region of reduced substrate resulting from consumption by the counter wave. The result is a mutual retardation of the waves. After collision, survival is prolonged since substrate diffusing from the surrounding regions provides additional support for the activation (the local stabilizing effect of diffusion was discussed in detail in Figure 4.4). Together, the effects of retardation and elongation lead to a blob-like structure or a sharpening at the tip of the V's (arrows in Figure 5.5 a, b).

Although very different in its appearance, another phenomena has presumably the same basis. Frequently a particular orientation of waves dominates because the waves in a small region that run in the opposite direction disappear (Figure 5.5 c, d). According to the model, the long-range antagonist causes a delay in the initiation of the short counter wave, the ∧-like structure. Due to this delay, the counter wave will become shorter and shorter over the course of time until the

Position →

Time →

Figure 5.5. Traces of the diffusible antagonist in *Bankivia fasciata*. (a) The V-shaped zones of annihilation show some sharpening at the tip. (b) Model: The action of the long-range antagonist can lead to a reduction in the speed of the waves shortly before collision. (c-d) Small irregularities on oblique lines disappear. (e) Model: the initiation of short counterwaves becomes delayed by the diffusible antagonist until no locally advanced trigger can take place. [S55, photographs kindly supplied by J. Campbell, see also Ermentrout et al., 1986].

dominating wave reaches the initiation point before the spontaneous initiation can take place (Figure 5.5e).

The shell in Figure 5.6a exhibits a pronounced chessboard pattern decaying into oblique lines. As outlined earlier, if the autocatalysis saturates, the activated and non-activated phase can be of the same length, causing thick parallel lines (see Figure 3.5). It is the second and diffusible antagonist that enforces the phase shift between parts of these lines. In the time record, this leads to the chessboard pattern. A higher substrate production causes the transition to oblique lines, a transition that is analogous to that shown in Figure 5.2c-b for the non-saturating case.

Even if the chessboard pattern resolves into oblique lines, some elements can remain that are reminiscent of this pattern. Either the initiation or the annihilation

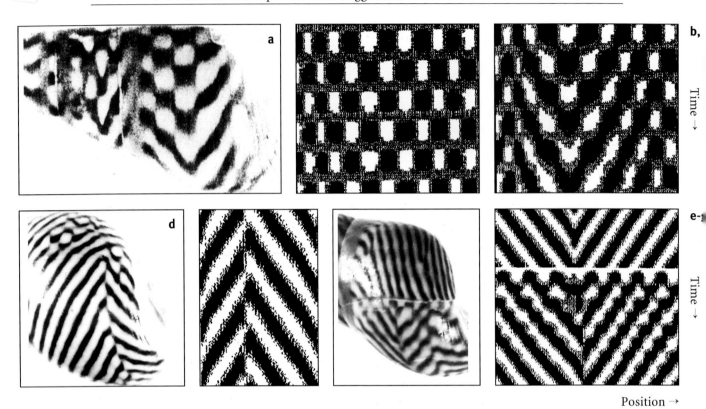

Figure 5.6. Chessboard pattern and its decay into oblique lines. (a) *Bankivia fasciata.* (b, c) Model: saturation of the activator autocatalysis leads to activation periods of the same length as the intervals in between. The highly diffusible antagonist (in this case, the substrate) causes the phase shift between parts of the resulting thick lines. Low inhibitor diffusion maintains the maximum phase difference and the coherence of the patches. An increase in substrate production leads to the tendency to form oblique lines. (d, e) Initiation and (f, g) termination of travelling waves may occur out of phase due to the long-range antagonist. In (f) a global perturbation took place, simulated in (g) by a temporary reduction of the activator concentration. The transient formation of a more meshwork-like pattern is reproduced. (Photographs kindly supplied by J. Campbell); [S56, S55].

of waves may occur out of phase. Isolated columns of alternating black/white and white/black transitions can remain stable over a long period of time (Figure 5.6d-g). According to the model, the effect of the long-range antagonist is especially pronounced if two waves that run in opposite directions are close to each other. This is the case either at the initiation or the termination of waves. Waves can be generated in alternating directions, one to the right, one to the left, etc. since one wave inhibits the other. The small group of cells from which the waves spread maintains a nearly steady state concentration.

The situation is similar during annihilation (Figure 5.6). The long-range antagonist stops the wave before a proper collision takes place. Again, the resulting black/white - white/black line can be regarded as a columnar fragment of the chessboard pattern. The chessboard and out-of-phase patterns require similar ranges of parameters. Especially important to this type of handshaking are activated and non-activated periods of about the same length.

5.5 Crossings and branching

The shells of some species are decorated with oblique lines that frequently cross each other. An example was given in Figure 5.1. Apparently, travelling waves can penetrate each other and enter into a region that should be refractory, in contrast to the normal rules for waves in excitable media. Survival of colliding waves is a feature of the meshwork patterns discussed above (Figure 5.3). However, patterns with general crossings are usually much less regular.

How can travelling waves survive a collision? One possibility is that during a collision cells remain activated until the refractory period of the neighbouring cells is over. Thereafter, a re-infection of these cells can take place. A crossing can be regarded as the initiation of a new pair of waves at the point of collision. A signal must be available that enables the cells to remain activated. What can the signal be? What is different at the point of collision? In the normal chain of triggering events that leads to oblique lines, the activation of one cell is followed by a full activation of its neighbouring cell. In contrast, after a collision the activation breaks down, and no subsequent activated cells are available. Thus, for a group of cells, a possible signal that a collision took place at this position could be the absence of subsequent activation.

We have discussed the possible role of a second rapidly diffusing antagonist and we will see that the same mechanism (Equation 5.1) can be used to simulate the formation of line crossings, although with a different range of parameters. Let us assume an activator-substrate system tuned to bi-stability, i.e., a low and a high steady state exist (see Figure 3.3), and that the activator is diffusible while the substrate is not. The resulting pattern would not be very interesting. From a local initiation, the activation would spread until all cells are switched from the low to the high steady state (Figure 5.7a). Now let us assume a highly diffusible inhibitor also exists. During the spread of activation, each newly activated cell also produces a burst of diffusible inhibitor which shifts the recently activated cells back into the non-activated steady state. Travelling waves result that are very similar to those generated by the elementary mechanism. However, such waves behave differently upon collision. There, the decline of activation also causes a decline of the rapidly diffusing inhibitor. This creates a counter regulation. If the time constant of the inhibitor is fast enough, this reduction will no longer allow the cells to shift from the high to the low state. The waves stops but a local activation remains. When the refractory period of the neighbouring cells is over, two new waves are initiated that travel in opposite directions in much the same way as waves that penetrate each other, except for a small delay at the point of collision. The inhibitor that emanates from the reactivated cells suppresses the activation in those cells in

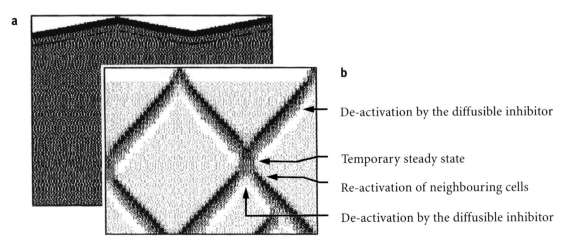

De-activation by the diffusible inhibitor

Temporary steady state

Re-activation of neighbouring cells

De-activation by the diffusible inhibitor

Figure 5.7. Formation of crossings by an additional diffusible inhibitor. (a) An activator-substrate model is tuned to bi-stability. High activation spreads due to activator diffusion. The high state is reached with some overshoot. (b) The additional diffusible inhibitor (red) causes each subsequently activated cell to down regulate the activation in previously activated cells. Apparently normal travelling waves result. However, at the point of collision the activation survives due to the rapid decrease of the additional inhibitor concentration. After recovery of the substrate (green), two new waves are initiated [S57, S57a].

which the activation has survived. Figure 5.7b shows a computer simulation and the presumed sequence of events.

A shell with crossings, as well as a simulation on a larger field, is shown in Figure 5.8. A comparison reveals that the model is able to account for fine details. Although as a rule both waves survive a collision, in some instances one of the new waves becomes extinct, causing a pattern like an amputated X. The initial attempt to form a second wave is clearly visible in most cases. In the model, most of the additional diffusible inhibitor is secreted after full activation of the two new waves. This leads to a competitive situation in which one of the two waves may not survive. Note that the competitive effect becomes critical only after a collision, not before, although in both cases the waves pass through a stage of comparable distances. After collision much less substrate is available for supporting the waves (see Figure 5.7). The situation is especially critical if another wave is nearby.

In addition to crossings, branching of the oblique lines also occurs in the shell shown in Figure 5.8, indicating that wave splitting occasionally takes place. At a particular time a secondary wave is triggered that moves in the opposite direction

Figure 5.8. Formation of crossings: Patterns on *Tapes literatus*. Simulation calculated with an activator-depleted substrate model plus an additional inhibitor (red). The waves can survive a collision. Occasionally wave splitting occurs, leading to a branch on the shell. The general pattern discontinuity visible on the shell is simulated by a reduction of the activator concentration to 20% of its current value in each cell. This leads simultaneously to the termination of waves and the initiation of new pairs of waves [S58; S57 provides a simulation without branchings].

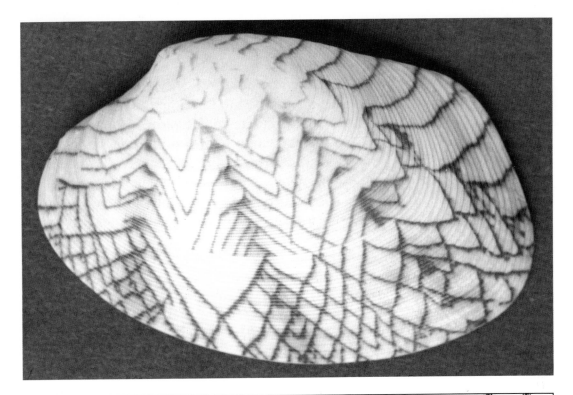

Time →

Position →

with the same speed. This indicates that inhibition can become insufficient to achieve a complete switch from the high to the low steady state.

The pattern on the shell shown in Figure 5.8 displays a global irregularity presumably caused by a sudden external event such as dryness or lack of food. Many waves have vanished at the same time. However, other waves survived and even formed new waves by wave splitting. The model resolves this apparently contradictory behaviour between the extinction of waves and the simultaneous generation of new waves. A temporary lowering of the activator can lead either to wave termination or, due to the concomitant reduction of inhibitor concentration, to the initialization of a new pair of waves, as in a normal collision. In some cases, the attempt to form two waves is clearly visible but one wave does not survive the mutual competition.

The details on the shell in Figure 5.8 also reveal an aspect not yet included in the model. Relief-like structures separated by faint lines parallel to the growing edge can be seen. Between these lines, the pigmentation increases in width, but at the line this width is discontinuously reduced. The pigmentation has, in fact, a fine structure resembling a chain of triangles. Thus, the shift from the high to low steady state is not, as assumed so far, a continuous process but depends on, or is enhanced by, an independent oscillatory process. One suggestion is that these fine lines result from an external periodic event such as a daily rhythm. However, sometimes two such lines merge, indicating that the synchronism of the oscillations is not absolute. Therefore, the periodic line formation must result from an internal oscillatory process. In other shells, the feature of chained triangles is more obvious (Figure 8.1) and we will come back to the underlying mechanism (Figure 8.5).

5.6   Changing the speed of the wave during a collision

Depending on how rapidly a system recovers after a collision, some time may be required before the two new waves can be emitted. The local temporary steady state after the collision leads to a thin line parallel to the direction of growth. Since the substrate concentration may not be fully recovered yet, the speed of the waves may be slow initially but may increase continuously until the next collision takes place. Figure 5.9 show this feature very clearly. The accompanying simulation was performed under the assumption that two substrates are available for maintaining

Figure 5.9. Changing wave speed during a collision: generating pattern elements reminiscent of staggered wine glasses. Shell of *Conus abbas* . One characteristic is the bending of the lines before collision indicating an increase in speed. After the collision some time may be required before a wave starts moving. The simulation was calculated with a two-substrate model (Equation 5.2). The actual mechanism, however, is presumably more complex and involves two complete pattern forming systems, see Figure 7.2 and 7.8 [S59].

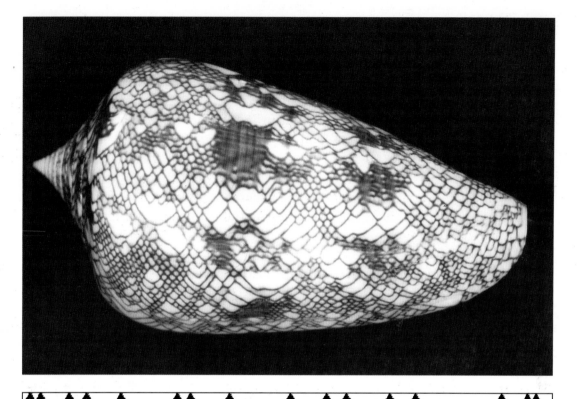

Time →

Position →

the autocatalysis (Equation 5.2). The non-diffusible substrate is responsible for the movement of the waves. The diffusible substrate enables their survival after a collision. The resulting pattern on the shells is reminiscent of staggered wine glasses. The rounded shape at the top is followed by a thin stem below. Later we will discuss evidence from related patterns indicating that not only a second antagonist but also an additional pattern-forming reaction is reponsible for the alternation between the pulse-wise and the steady state activation (Chapter 7).

On the shell shown in Figure 5.9, regions with different mesh size are visible, generated either by the initiation of an additional branch or by the failure of the waves to survive a collision. A definite mesh size can evolve diagonally in a neighbouring region. This feature is clearly reproduced in the simulation.

5.7 Parallel and oblique rows of staggered dots

Another frequent pattern with a periodicity in space and time consists of rows of staggered dots. Activation in discrete patches suggests that both antagonists diffuse. This can cause patch-like activations separated along the space coordinate as well as along the time coordinate. This pattern occurs with a wide range of parameters if both antagonists are inhibitors (Equation 5.3). As mentioned, when an inhibitor is involved new maxima are inserted at some distance from existing activated regions (see Figure 2.7). This is precisely what happens in the staggered dot patterns, except that the maxima are no longer stable but disappear after a short time due to the second antagonist. In most of the simulations shown so far in this chapter, one antagonist has had a high diffusion rate, the other was non-diffusible. However, when the second antagonist has a moderate diffusion rate, travelling waves are no longer possible since even this slowly diffusing antagonist spreads faster than the activator. Under these conditions and with the two antagonists having a time constant longer than the activator, activation occurs in separated patches. (If both antagonists were highly diffusible, stable stripes would result, as shown in Figures 2.3 and 3.5).

On some shells the dots appear to be arranged in oblique lines. On others the dots are in rows parallel to the growing edge but in staggered positions. The simulations in Figure 5.10 illustrate the basis of this effect. Natural patterns of both types are given in Figure 5.11. Simulations have been performed with the two-inhibitor model. They differ only in the range of the slowly diffusing inhibitor which forms a forbidden zone after each activation (green in Figure 5.10). If its diffusion is more widespread, the subsequent activation will be further from the previous activation. But this distance cannot be larger than half the distance between two previous activations since an activation cannot take place in the forbidden zone of a neighbouring activation. Therefore, from a certain spacing onwards, subsequent activations will occur precisely between two preceding activations. This leads to a very regular pattern of dots out of phase. In addition to the regular spacing, a

Equation 5.3 and 5.4: Extensions
of the activator-inhibitor mechanism

A second inhibitor $c(x)$ is assumed to be involved in a normal activator-inhibitor interaction (Equation 2.1). The second inhibitor may have an influence on the production of the activator as well as on the production of the primary inhibitor b (x):

$$\frac{\partial a}{\partial t} = \frac{s}{c}\left(\frac{a^2}{b} + b_a\right) - r_a a + D_a \frac{\partial^2 a}{\partial x^2} \tag{5.3.a}$$

$$\frac{\partial b}{\partial t} = \frac{r_b\ a^2}{c} - r_b\ b + D_b \frac{\partial^2 b}{\partial x^2} + b_b \tag{5.3.b}$$

$$\frac{\partial c}{\partial t} = r_c(a - c) + D_c \frac{\partial^2 c}{\partial x^2} \tag{5.3.c}$$

The two antagonists are not completely equivalent. While the primary antagonist b has a direct influence on activator production but not on itself, the second antagonist also slows down the rate of inhibitor production. Similar to the description in Equation 5.1, this has the consequence that a is independent of c over a wide range. Therefore, the second inhibitor changes the region that will ultimately become activated but does not primarily change the absolute activator concentration.

Alternatively, the second inhibitor may have an additive inhibitory effect:

$$\frac{\partial a}{\partial t} = \frac{s\ (a^2 + b_a)}{(s_b b + s_c c)} - r_a a + D_a \frac{\partial^2 a}{\partial x^2} \tag{5.4.a}$$

$$\frac{\partial b}{\partial t} = r_b(a^2 - b) + D_b \frac{\partial^2 b}{\partial x^2} + b_b \tag{5.4.b}$$

$$\frac{\partial c}{\partial t} = r_c(a - c) + D_c \frac{\partial^2 c}{\partial x^2} \tag{5.4.c}$$

In this case, increasing the c concentration chokes the activation. After localized activation, the spread of activation soon stops. Triangle-shaped patterns will result (Figure 5.12.)

synchronization in time also takes place. Since the two new activations have the same distance from the previous spots of activation, the lowering of the highly diffusible inhibitor is symmetric and the triggering of subsequent activations occurs simultaneously. Thus, neighbouring dots appear on rows more or less parallel to the axis although a slight bending of these rows indicates that some phase shift may accumulate over the total length of the mantle gland.

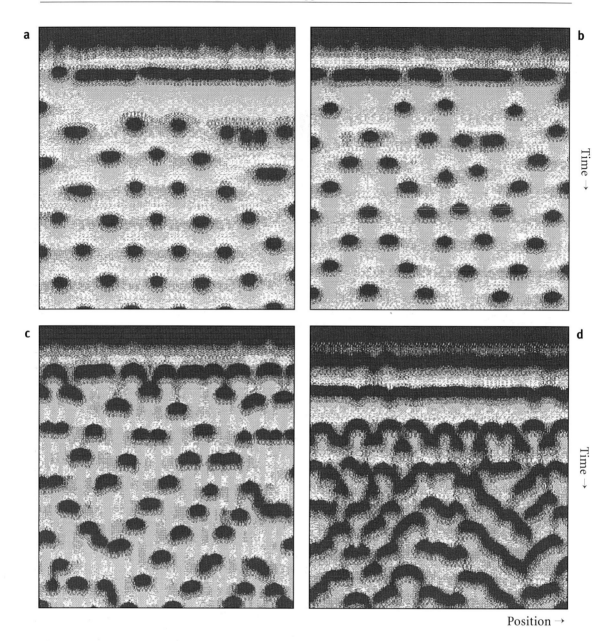

Figure 5.10. Horizontal and oblique rows of dots. Separate patches of activation occur if both antagonists diffuse. The simulation was calculated using the two-inhibitor model, Equation 5.3. (a) If the rate of the slowly diffusing inhibitor (green) is above a critical level, the patches appear in horizontal rows and at positions precisely out of phase. (b) With a lower diffusion rate, the patches appear in oblique rows. (c) Without diffusion the patches partially merge. (d) When the low or non-diffusible antagonist has a shorter lifespan, a transition from patches to oblique lines occurs [S510].

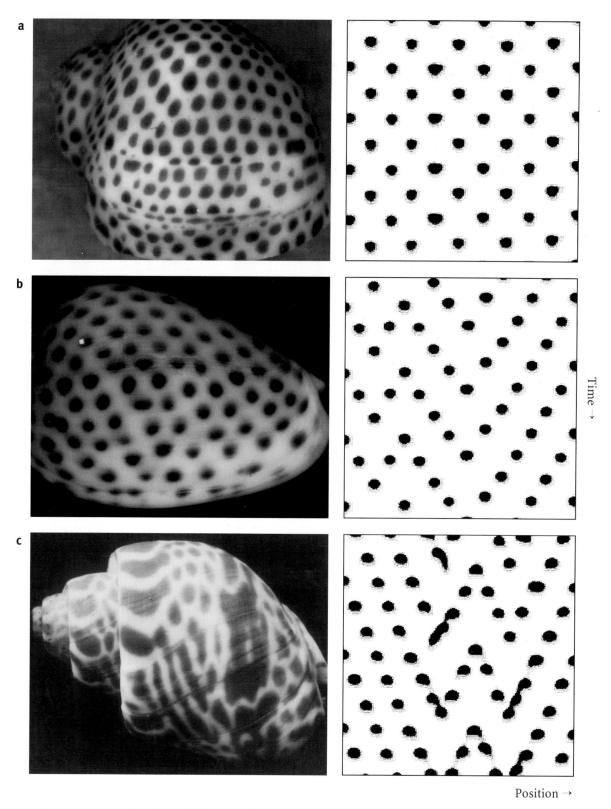

Figure 5.11. (a-b) Minor differences decide whether dots are arranged in horizontal or oblique rows. (Shells of *Natica stercusmuscarum* and *Persicula persicula*) (c) Shell of *Babylonia japonica*. The simulation reproduces the transition from isolated patches to oblique lines [S510].

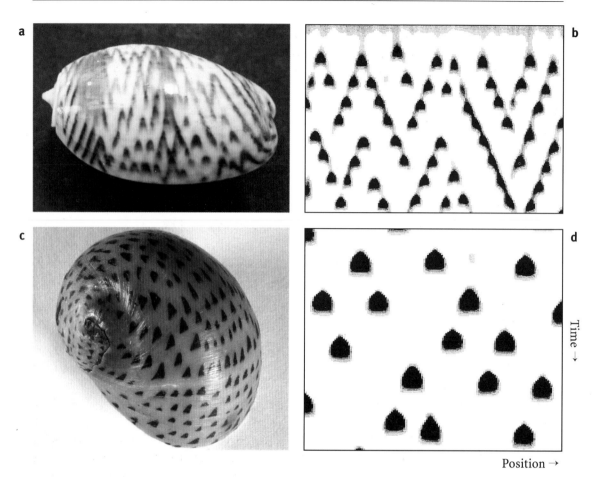

Figure 5.12. Crescents and triangles are the preferential pattern if two inhibitors act in an additive way. (a) Shell of *Oliva bulbosa*. (b) If the primary inhibitor is non-diffusible, the crescents can be connected by faint lines resulting from waves of low level activation. (c) Shell of *Neritina penata*. (d) In contrast, when the primary inhibitor diffuses at a specific rate, a separation of the patches occurs and isolated triangles are formed [S512].

In contrast, if the rate of the slowly diffusing inhibitor is lower, a pattern with patches arranged along oblique lines will emerge. Successive patches of activation appear with a slight offset in space, but only on one side of the forbidden zone. An isolated activated patch gives rise to two diverging rows and two merging lines disappear, analogous to the behaviour of travelling waves. At even lower diffusion rates, or without diffusion, the patches may fuse to become partially connected oblique lines. Figure 5.11 provides examples of the different modes of dot patterns. The range of parameters over which one or the other pattern is possible will overlap. In a critical range, both types may be formed. The system can lock into one or the other mode. Both types may even appear in different regions of the same shell and one mode may replace the other over the course of time. Therefore, neither pattern is exclusive.

On some shells, the patches have a triangular or crescent-like shape. Figure 5.12 shows two examples. According to the model, this pattern emerges preferentially if the two inhibitors work in an additive way (Equation 5.4). The inhibitor with the long time constant poises the activation in such a way that it terminates. Similar to the case with dots (Figure 5.11), a non-diffusible primary inhibitor allows a partial connection of the crescents (Figure 5.12a). The pattern has many of the same features as those discussed above. Patches become partially connected to oblique lines that can cross or branch. Frequently, one or both waves terminate before a collision takes place, supporting the postulated feature of lateral inhibition. One remarkable feature is that the faint pigmentation lines that connect the crescent-like patches have the same inclination as the heavily pigmented regions. This seems to be counter-intuitive since cells with a lower activation should require more time to infect their neighbours.

5.8 Conclusion

By a second antagonist an excitable system consisting of many coupled oscillators may obtain unusual properties. Travelling waves can occur that penetrate each other, causing crossings in the shell patterns. Wave splitting and thus branching is possible as well. The second antagonist can enforce a de-synchronization of the oscillators. This is much in contrast with the behaviour of many other coupled oscillators in biology that tend to synchronize. Flashing fireflies, the pace-maker cells of the heart, networks of neurons in the circadian pace-maker, and the insulin-secreting cells of the pancreas are examples for synchronizing systems. It has been postulated that this synchronization is a general feature of pulse-coupled oscillators (Mirollo and Strogatz, 1990; see also Stewart, 1991). As the staggered dot patterns on shells demonstrate, this rule is not universal. It is violated when several agents are involved in the coupling and especially if the coupling is not only achieved by the exchange of activating but also of inhibiting signals. On the other hand, waves with solitary properties, i.e. waves that penetrate each other and that may split, have been also observed in other systems, for instance on waves of catalytic oxidation of Carbon oxide at the surface of Platinum crystals (Bär *et al.*, 1992). These authors assumed imperfection on the crystal structure as the cause of this phenomenon. The systematic occurrence of penetration and splitting of waves in shell pigmentation precludes an analogous explanation. But, the other way round, the shell patterns suggest to consider long range inhibitory effects as an essential ingredient whenever waves with solitary properties are observed. Thus, shell patterns may contribute to a better understanding of other dynamic systems.

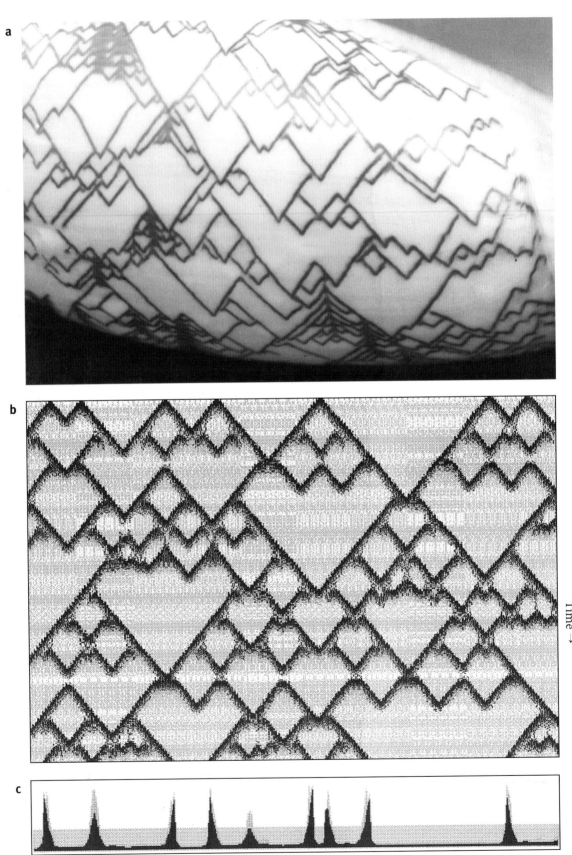

Chapter 6

Branch initiation by global control

6.1 Simultaneous pattern change in distant regions

In the mechanisms discussed so far, only information exchange between neighbouring cells has been considered. However, several patterns indicate that particular events occur simultaneously at very distant positions. For instance, the shell of *Oliva porphyria* (Figure 6.1) shows oblique lines with branching. A branch along an oblique line indicates the sudden formation of a backward wave. Many branches are initiated at the same time at distant locations. The dashed line in Figure 6.2b indicates such a moment. There would not be enough time for a signal to travel over such a long distance by diffusion. On the other hand, not all lines branch at this critical moment. This indicates that from time to time a signal is generated everywhere in the mantle gland that greatly enhances the *probability* of wave splitting. Such a signal must be the result of a global control.

By analogy with other secretory organs, Ermentrout et al. (1986) proposed a neuronal basis for controlling pigment secretion. *Via* neurons, a signal could be transmitted very fast. So far, direct evidence for neuronal control is lacking. As an alternative, we will assume that the global signal results from a hormone-like substance that circulates within the animal. In this way the signal is simultaneously available throughout the animal (Meinhardt and Klingler, 1987, 1991). This chapter will discuss the types of couplings that lead to global control of branch initiation. It is assumed that the signal is produced by the patterning system itself. Chapter 8 will discuss systems in which global signals result from an independent oscillating process.

Figure 6.1. (a) Shell of *Oliva porphyria*. Branchings occur simultaneously at distant positions. (b) In the model, a homogeneously distributed hormone (green) is assumed that mediates branching. This occurs whenever the number of waves and, therefore, the hormone concentration becomes too low (light green regions; see Figure 6.2 for details). (c) Distribution of the substances at the end of the simulation shown in (b). [S61; S61a displays the distributions shown in (c) in a movie-like manner].

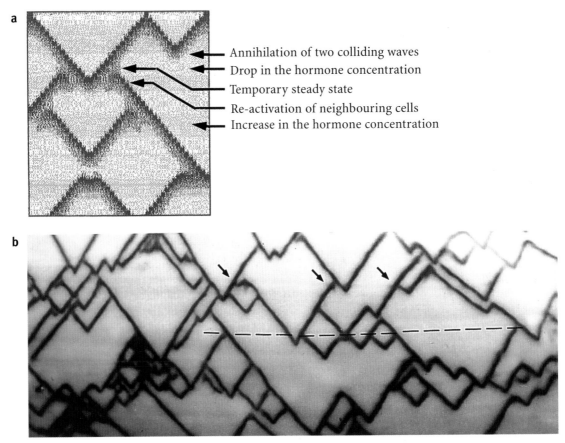

a

Annihilation of two colliding waves
Drop in the hormone concentration
Temporary steady state
Re-activation of neighbouring cells
Increase in the hormone concentration

b

Figure 6.2. Model for branch formation. A branch indicates the sudden formation of a backward wave. In this model branching occurs whenever the number of travelling waves drops below a certain threshold value. The controlling agent is a hormone-like substance c (green) that controls inhibitor lifetime. It is produced at a rate proportional to the local activator concentration. Its rapid distribution is simulated by averaging. Below a certain c concentration (light green) the activated cells switch from an excitable to a steady state mode of activator production. Groups of cells remain activated longer than the refractory period of their neighbouring cells. This initiates the backward waves. Since c is uniformly distributed, many oblique lines branch simultaneously. (b) Detail from Figure 6.1. The dashed line marks a moment where many branch formations took place. The arrows mark a temporary increased background pigmentation.

6.2 Branch formation in *Oliva porphyria*

From the point of view of the normal behaviour of waves in an excitable media, the sudden formation of backward waves and branching is a very unusual event. The new waves must spread into a region that should be refractory due to the primary wave.

A closer inspection of Figures 6.1a or 6.2b indicates a prerequisite for branch formation: the refractory period must be very short. Immediately after branching, the new oblique lines have the same inclination, indicating that the waves start with the same speed. The system must return to its former degree of excitability very fast. Another feature points in the same direction. Two oblique lines can be very close together, indicating that one wave can be followed by a second wave after a very short time interval. On the other hand, the straightness of the oblique lines indicates that excitability does not increase further if cells remain non-activated for a long period of time.

The formation of a branch requires, for a short period, a modification of the simple serial triggering mechanism that generates oblique lines. A particular cell, or group of cells, has to remain activated for a prolonged period so that its activation "survives" the (short) refractory period of its earlier activated neighbours. After this period a re-infection of previously activated neighbouring cells is possible, thus initiating the backward wave. A splitting of the wave has occurred. Therefore, the formation of branches requires a signal that causes an elongation of the activation period. What could the signal be? Usually, if two travelling waves collide, both become extinct. The ∨-like pattern elements in Figure 6.1 show the protocol for this annihilation. With each collision the number of travelling waves along the growing edge becomes smaller. One way to compensate for this loss is to form new lines by wave splitting. This suggests that the signal for branch formation is triggered whenever the number of travelling waves becomes lower than a certain threshold. The total number of activated cells at any given time provides a measure of how many travelling waves are present. If each activated (pigment-producing) cell produces a hormone-like substance c that is rapidly distributed over the total number of cells n, the c concentration is proportional to the ratio of activated and non-activated cells.

To initiate branch formation, the cells must switch from burst-like activation to steady state activation whenever the c concentration drops below a critical level. When the refractory period of a neighbouring cell is over, a backward infection becomes possible. When such global branch formation occurs, the number of travelling waves doubles. This also causes a doubling of the hormone concentration c. All cells switch back to the excitable mode of activation. Thereafter, all oblique lines, including those just initiated by wave splitting, are elongated by the normal chain of triggering events.

How can the c concentration determine whether the activation is in a steady-state or a burst-like mode? In the activator-inhibitor model (Equation 2.1b), pulse-

Equation 6.1: Branch formation by changing time constants via a hormone-like substance

An activator-inhibitor system is assumed (Equation 2.1). In addition, the half-life of the inhibitor assumed to be under control of a hormone-like substance. Each activated (pigment-producing) cell produces the hormone c that becomes rapidly distributed within the total number of cells n.

$$\frac{\partial c}{\partial t} = r_c \sum_{i=1}^{n} a_i/n - r_c c$$

All activated cells contribute to hormone production. The $1/n$ term results from the homogeneous distribution of the hormone c among all cells n. Due to this averaging, the c-concentration is the same in all cells along the growing edge and decreases uniformly with a decreasing number of a-producing cells, i.e., with the number of travelling waves.

The hormone is assumed to modify the stability of the inhibitor. The decay rate of the inhibitor r_b in Equation 2.1b has to be replaced by the effective decay rate of the inhibitor

$$r_{b_{eff}} = r_b/c$$

If the total number of travelling waves becomes too small, the inhibitor lifetime will become so short that the activated cells switch from oscillating to steady state activation. This enables wave splitting.

An elongation of activator lifetime ($r_{a_{eff}} = c\, r_a$) or an increase in substrate production ($b_{b_{eff}} = c\, b_b$) using an activator-substrate system is possible as well. However, an examination of the details of the resulting pattern exclude these possibilities (see Figure 6.7).

Time →

Position →

Figure 6.3. Pattern heterogeneity on the shell of *Oliva porphyria* L. and a simulation on a large array of cells. Depending on random fluctuations and the development of the system, branches may occur at particular locations in rapid succession such that a dense cluster of lines results. In the simulation this pattern occurs only when the field size is large [S63; depending of the random fluctuations the actual simulation may look quite different].

like activations occur if inhibitor decay is slower than activator decay, i.e. if $r_b < r_a$, but remains in a steady state when $r_b > r_a$ (see Figure 3.2). If the inhibitor is stabilized by the hormone, a lower c concentration leads to more rapid inhibitor decay. In the correct range this will lead to a shift from pulse-like activator production to steady state and thus to branch formation. The substitution of r_b by r_b/c has such an effect (see Equation 6.1). To reproduce the pattern of *Oliva porphyria* a further requirement is that cells become activated only by activated neighbours, not spontaneously. Cells must be excitable, but not oscillating. This is the case when a basic inhibitor production b_b or a high Michaelis-Menten constant s_b is given (Equation 3.1 and Figure 3.2). Figure 6.2 illustrates the formation of a single branch, while Figures 6.1 and 6.3 show simulations with a larger field of cells.

The simulations reproduce many details of the natural pattern. The high probability of simultaneous initiation leads to many ∨-like elements with the same distance between the tip and the branch initiation points. While the original wave proceeds undisturbed, a branch frequently appears only loosely connected to the original line. In the simulation, this effect results from an overshoot during the normal chain of triggering events which is much higher than the temporary steady state activation (see Figure 3.2). Sometimes, a small hook appears close to the point of branch initiation. This is the result of an incipient wave initiated parallel to the primary wave. Usually this wave does not survive since the cells are not yet excitable enough. If this wave does survive, however, a parallel line close to the original line is formed. Very dense local branching can occur if, by accident, many waves become extinct almost simultaneously. The low density of waves along one section of the growing edge causes an increase in branching along another section. The result is a plethora of pigment lines in one area (Figure 6.3).

6.3 Pattern diversity

Simulations of *Oliva porphyria* can be used to demonstrate the origin of pattern diversity within the same species. According to the model, minor differences can be decisive in branch formation. Once made, such a decision has, of course, a dramatic effect on subsequent decisions. Therefore, even if two systems are initially in almost identical situations, after a few generations of branch formation all indications of this are lost. This is the well known behaviour of chaotic systems. Therefore, unavoidable minor differences are sufficient to generate a diversity of patterns. To illustrate this feature, Figure 6.4 shows two simulations plotted on top of each other in different colours. Either random fluctuations or slightly different initial conditions have been assumed. Initially both patterns are nearly identical, the red pattern almost completely covers the black pattern. With each decision to form or not to form a branch, the patterns becomes more different and the differences spread to neighbouring regions.

6.4 The influence of parameters

The overall appearance of patterns generated in this way depends on several factors. According to the model an equilibrium is maintained between the generation of new oblique lines by branching and their pairwise termination by collisions. More important for the appearance of the pattern is the number of waves required to attain this equilibrium. This number depends on the hormone-controlled decay rate of the inhibitor. If this parameter is large, many waves are formed and a dense network of pigmented lines emerges. On the other hand, if the parameter is small, individual oblique lines are the dominating pattern element. Figure 6.5a shows the transition from high to low wave density generated by a change in this parameter.

a

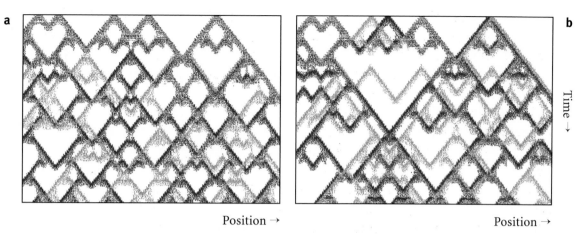

b

Time →

Position →   Position →

Figure 6.4. Generation of pattern diversity. In order to show that minute differences lead to different patterns two simulations are plotted on top of each other (black and red). (a) The same cells are activated initially but different random fluctuations exist. At early stages (top of the figure) both patterns are very similar; the black pattern is largely covered by the red. Small differences decide whether branch formation occurs. After a few generations of branching, the patterns are completely different. (b) Two calculations with the same random fluctuations. Some cells are activated at different positions initially [GT64].

If the time constant of the hormone is much larger than that of the activator, very dense bundles of oblique lines are formed preferentially that run parallel to each other. Many branches will then be formed in rapid succession, since a relatively long period of time is needed before the hormone concentration adapts to the elevated density of the waves. Conversely, it may require a longer period of time before the next set of branches can occur. Figures 6.5b,c show simulations of low and high wave density. At still higher densities, a transition into triangles may occur. Although, as will be discussed in chapter 8, triangles are a frequent pattern, it is unlikely that they are formed by this mechanism. In the simulation in Figure 6.5d it can be seen that the upper tips of these triangles are in a permanent steady state. With an increasing number of activated cells a continuous shift towards oscillations takes place. This is the reason for the stripe-like elements along the lower border of each triangle. However, when real shells have triangles with a stripe-like fine substructure, they do not show this systematic evolution from tip to base (see Figure 8.10).

On some shells with global control, travelling waves occasionally die out even when no collision takes place. Events of this type are seen in many locations on the shell shown in Figure 6.6. In the model, this occurs when a relatively high activator concentration is required to maintain the chain of activation. As mentioned in the activator-inhibitor mechanism such a high threshold results from a high Michaelis-Menten constant s_b (Equation 3.1) or higher basic inhibitor production b_b (Equation 2.1b). If the activation becomes too low it returns to the low steady state and the line terminates.

Figure 6.5. Types of patterns that can be generated by differing the parameter for hormone-like control. The decay term of the inhibitor r_b determines the density of branching. (a) A change in the pattern caused by lowering r_b by a factor of 3. (b) A hormone with a larger time constant leads to a cluster of branches with longer, non-branching lines in between. (c) As (b) with higher density. (d) At even higher densities, triangles may be formed that can disintegrate into a pattern of dense lines. This transition may spontaneously revert back to triangles later on [S65a-S65d].

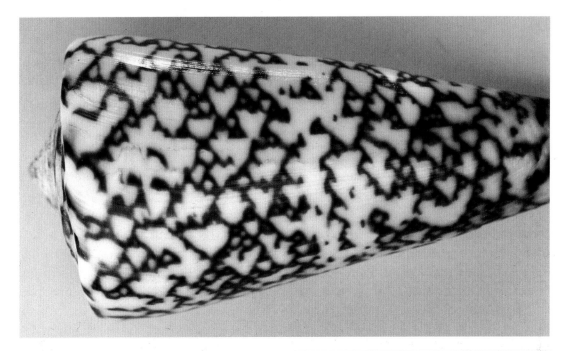

Time →

Position →

Figure 6.6. Global control and line termination: shell of *Conus thalassiarchus*. In the model, when high activator concentration is required to trigger neighbouring cells, oblique lines can also terminate without collision. Since, on average, termination takes place after a certain time interval, the wave cannot proceed very far. Regions appear spontaneously which do not form pigmentation for longer time intervals [S66].

6.5 Alternative mechanisms

A change in the lifetime of the inhibitor under hormone control is, of course, only one possible mechanism to achieve the switch back and forth between burst-like and steady state activation. An elongation of activator half-life or an increase in substrate production using an activator-substrate model would be appropriate as well (Figure 6.7). However, the details of the resulting patterns look very different from the natural pattern, making such a mechanism unlikely. An elongation of activator lifetime leads to a characteristic thickening of the line shortly before and during branch formation, at variance with the natural pattern.

As mentioned, the fact that branches have the same inclination as the original wave indicates that the system returns to normal excitability very fast. In a depletion mechanism this would require very high substrate production *and* a high decay rate independent of the autocatalysis (r_b in Equation 2.4). These conditions appear unrealistic since they would require such a waste of energy just to produce and destroy molecules. However, if these conditions are not met, the waves that follow a primary wave would spread at a lower speed due to substrate consumption by the forerunning waves (Figure 6.7b). The resulting line would appear steeper and bent. No such features can be seen on the natural pattern.

On the other hand, the actual mechanism may be more complicated. Since no lateral inhibition is assumed, the mechanism has no pattern forming capability on its own. The generation of a pattern requires initiation at particular positions. This restriction would be resolved by a second long-range antagonist such as that

Position →

Time →

Figure 6.7. Alternative mechanisms for branch formation. (a) The hormone reduces the decay rate of the activator. The pigment lines show thickenings shortly before branching takes place (arrow). (b) The hormone increases substrate production. Waves that follow shortly behind a primary wave are retarded due to substrate depletion. The resulting lines are finer and steeper (arrow). Both features are absent on the natural pattern. (c) If a second long-range antagonist is involved, the system is capable of forming the proper pattern starting from homogeneous initial conditions. In addition, spontaneous initiation of pairs of waves (arrow) becomes possible [S67a-S67c].

described in the previous chapter. With this extension, spontaneous initiation of new pairs of diverging lines also becomes possible (arrow in Figure 6.7c). This is a feature that evidently exists in natural patterns (Figure 6.1). Since, on the one hand, very rapid re-establishment of normal excitability must be attained and, on the other hand, a spontaneous triggering is only possible after a very long time interval, the assumption that two inhibitory actions exist with very different time constants is reasonable. According to the models, this takes place whenever the concentration of the second long-range inhibitor becomes sufficiently low.

The list of possible mechanisms that may cause branching under global control is certainly not exhaustive. The mechanisms outlined above should be regarded only as elements of a tool box for reconstructing shell patterns. Essential features are that (i) branch formation can be initiated by a temporary change in the time constant of one substance, and (ii) a simple feedback of the density of travelling waves on the lifetime of the antagonist can lead to branching whenever the number of waves becomes too low. The result is an equilibrium between newly formed lines due to branching and the disappearance of waves after a collision.

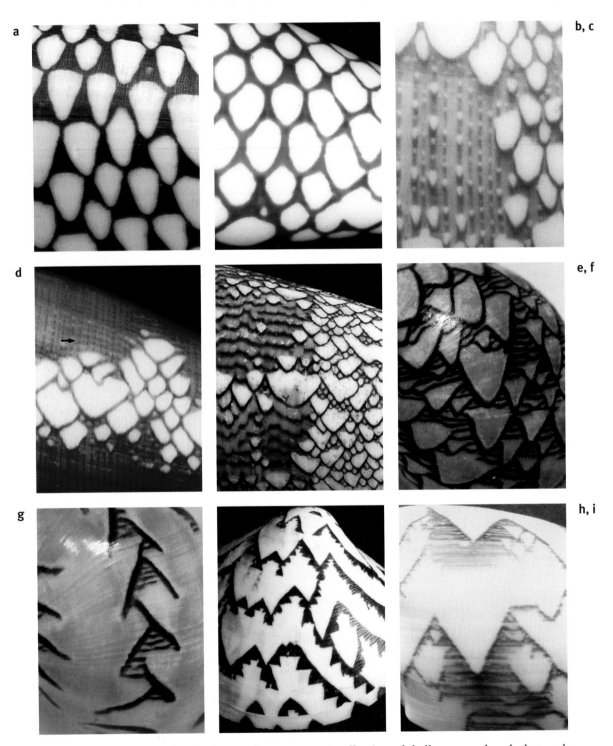

Figure 7.1. Inherent similarities in complex patterns. A collection of shells arranged such that each subsequent pattern contains elements of the preceding pattern.

Chapter 7

The big problem: two or more time-dependent patterns that cause interference

7.1   Inherent similarities in complex patterns

Many shells show patterns far more complex than those simulated so far. Figure 7.1 contains a collection of typical complex shell patterns. To show their inherent similarities, they are arranged such that each subsequent pattern contains elements of the preceding pattern as well as new features. *Conus marmoreus* (Figure 7.1a) shows white drop-like regions on a dark pigmented background. In *Conus nobilis marchionatus* (Figure 7.1b) the white drops are enlarged at the expense of the pigmented regions. The pattern is reminiscent of staggered wine glasses. *Conus pennaceus* (Figure 7.1c) shows, in addition, dark lines on a pigmented background, occasionally interrupted by small white drops. In *Conus auratus* (Figure 7.1d) the dark lines are maintained without the white drops. Instead, a periodic large-scale transition to oblique lines with crossings occurs. Shortly before this transition the continuous background resolves into narrow lines parallel to the growing edge (arrow). Such axially oriented parallel lines on top of a pigmented background are a characteristic pattern element in *Conus textile* (Figure 7.1e). Unpigmented regions with a drop-like shape occasionally appear. Their lower borders are formed by a dark pigmented line. In regions without pigmented background the pattern displays fine lines with wine glass shape as mentioned above. Similar parallel lines with occasional loops (tongues) are characteristic of *Clithon* (or *Neritina*) *oualaniensis* (Figure 7.1f). Here, however, the regions with fine lines are missing. In the specimen of *Clithon* shown in Figure 7.1g narrow-spaced parallel lines are framed by oblique lines, causing the overall impression of connected triangles, a pattern that is characteristic of the bivalved mussel *Lioconcha castrensis* (Figure 7.1h). Similarly, in the shell of the snail *Voluta vespertilio* (Figure 7.1i) very fine but short parallel lines give the impression of oblique lines. The triangles that are occasionally formed may become starting points for branches.

The hidden similarities in these patterns that overtly look very different suggest that only a few basic mechanisms are at work. The diversity must result from different parameter values or minor modifications. The perfect model would describe all these different patterns with a single set of differential equations. This is not yet possible, but a description of the basic elements and how they may arise will be given. The sequence as given in Figure 7.1 will be used as a guide for further analysis in this and the two subsequent chapters. A close examination of the shells will provide essential hints.

What can be the basis of this complexity? Chapter 4 discussed patterns resulting from modifications to the pigment-producing system by a second, not directly visible, pattern. These cases were easy to see through since the pattern that accomplished the modifications remained essentially unchanged over the course of time. This unchanging pattern is certainly only a special case since variation in a single time constant is sufficient to switch from a stable to an oscillating pattern. Therefore, it is assumed that the complex patterns described in this and the following two chapters result from a pigment producing system that is modified by one or more time-dependent, usually invisible, patterns.

To account for the different shell patterns that were governed by a stable modifying pattern, it was necessary to assume different modes of interference. Some shells indicated an enhancing or stabilizing effect, causing a locally higher oscillation frequency (Figures 4.2 and 4.5) or even a transition into steady state pigment production (Figure 4.10). Other shells indicated an invisible pattern that had a suppressing influence (Figures 4.12 and 4.13). As will be shown in this chapter, similar modes of interference are required for time-dependent modifying patterns.

The deciphering of complex patterns is extremely difficult. We do not know what the modifying pattern looks like, whether its influence is positive or negative, or which component of the modifying system interferes with which component of the pigment producing system - it could be either the self-enhancing or the antagonistic element. Does the modifying pattern change the production, the destruction, or the diffusion rate? Does the pigment-producing system, in turn, produce feedback for the modifying system? An enormous number of possibilities exist. And even if the correct mechanism is found, it is still difficult to find the range of parameter values in which the mechanism produces the expected patterns.

The similarities between complex patterns are very helpful in narrowing down the number of possible mechanisms. If two models describe a particular shell equally well but only one can also account for a related pattern, a correct decision can likely be made. Simulations of some apparently plausible models will be provided in which the deviation between the simulated and the natural pattern will indicate that the model can be ruled out.

On some shells, two distinctly different intensities of pigmentation are visible. These different pigmentation levels are presumably a direct expression of a modifying pattern and provide hints about its character. Thus, modulated background pigmentation can be a key in understanding the complexity of a pattern.

Modelling these complex patterns involves a much higher degree of uncertainty than the simulations given earlier. The models described here should only be regarded as a first attempt. It may require more words to describe the unsolved problems and the "if's" and "but's" of the models than the models themselves.

This may make the reading of this chapter more cumbersome. On the other hand it will show that many open problems still exist and should be taken as a challenge to find different and better solutions.

7.2 White unpigmented drop-like pattern on a pigmented background

On *Conus marmoreus* (Figure 7.2a), it is difficult to determine what is pattern and what is background. Are white drop-like patterns formed on a dark pigmented background or are somewhat rounded dark pigmented triangles formed that are connected to each other. A decision in favour of the latter possibility is suggested by considering the related pattern on *Conus nobilis marchionatus* (Figure 7.2b). There, the red pigmented lines are much finer indicating that travelling waves play an important role.

The oblique lines discussed so far (chapter 3 and 5) had a constant thickness. This told us that the time interval in which cells produce pigment remained constant during wave propagation. The switching off of activation spread with the same speed as its onset. In contrast, on *Conus marmoreus*, the spread of pigmentation proceeds relatively slowly while the termination of pigment deposition occurs almost simultaneously over a large region. It is, so to speak, a collective breakdown. Pigmented areas of nearly triangular shape result. The switching off occurs so abruptly and over such an extended region that there would be no time for a signal to spread by diffusion. Therefore, the signal for collective breakdown must be a sudden event at the end of a long preparation phase taking place during pigment production.

Some other features shed light on the underlying mechanism. Although pigment termination takes place simultaneously over a large region, this dramatic event obviously has no influence on the spread of activation. In other words, the cells become pigmented at the same rate regardless of whether a collective breakdown takes place in the neighbourhood. The breakdown only influences transition from an activated to a non-activated state, but not *vice versa*. Pigment deposition survives collective breakdown at the wave front. After a certain time interval pigment production becomes stable again and starts to spread in both directions.

7.3 Evidence of a sudden extinguishing reaction

The pattern on *Conus marmoreus* (Figure 7.2a) can be simulated under the assumption that two pattern forming reactions are superimposed (Figure 7.2c). One system produces the proper pigment and is characterized by a bistable mode of activation that spreads slowly. A second system extinguishes the first system when the first system has been in an activated state for some time. In Figure 7.2c this second system is shown in red. The red system is assumed to shorten the activator lifetime

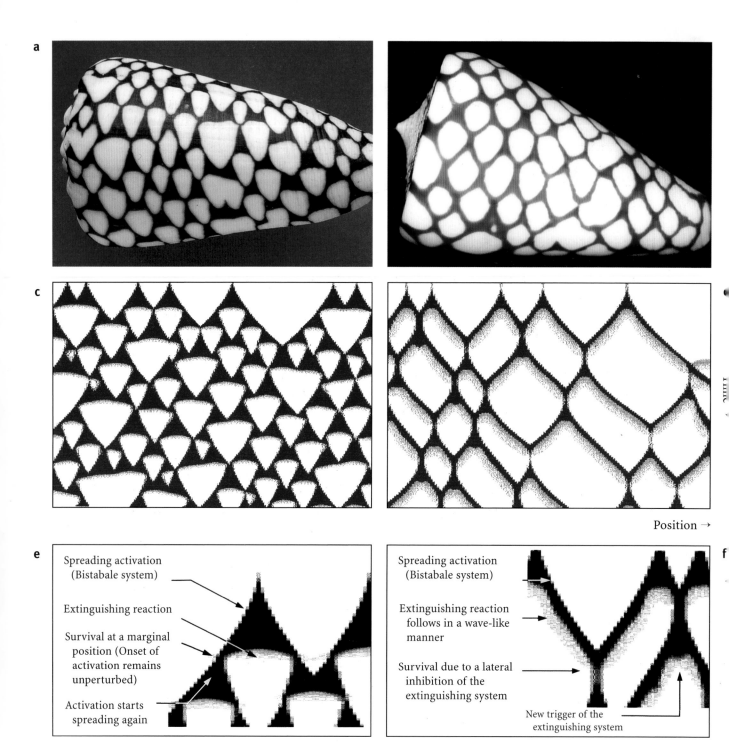

Figure 7.2. Formation of drop-like non-pigmented regions. (a) Pattern on *Conus marmoreus* and (b) *Conus nobilis marchionatus*. (c, d) Model: a bistable pigment producing system (black) exists. The activated state spreads slowly by infection. A second fast-spreading reaction (red) is triggered by the first one and causes its extinction. In (c), pigmentation must be produced for a certain period of time before the extinguishing reaction is triggered; in (d), this reaction can follow in a wave-like manner. Since the antagonist in the second reaction is diffusible, two extinguishing waves will not collide and activation of the first system will survive at the points of collision. (e, f) Details of the mechanism are described. These simulations are calculated using equations 7.1 a,b and e, f; (see page 111) [S72c, S72d]. The more complex patterns discussed later in this chapter indicate that local survival is an active process.

of the pigment producing system. Therefore, whenever the second system becomes active, the pigment system is transformed from a bistable to a pulse-wise mode of activation. In order for pigment termination to take place almost simultaneously over a large region, the components of the extinguishing system are assumed to spread much faster than those of the pigment system. Due to the slow spread of pigmentation and large scale termination, pigmented regions obtain a roughly triangular shape. The substrate in the extinguishing reaction is assumed to be produced only if a cell produces pigment. This causes the necessary delay between the onset of pigmentation and the trigger for the extinguishing reaction. It also enables the survival of the wave at the wave front where substrate production began only shortly before the extinguishing reaction was triggered. The substrate concentration for the extinguishing reaction is, therefore, lower and the reaction is less strong at this point. After the extinguishing pulse is over, pigmentation spreads from these zones of survival until the next extinguishing reaction is triggered. The pattern of black triangles and white drops results from the permanent interplay between the spread of activation, delayed large-scale extinguishing, and survival at the margins. The presumed sequence of events is shown schematically in Figure 7.2e.

The rapid destruction of the activator by the extinguishing reaction satisfies another requirement of the pattern: the absence of spontaneous re-activation. Because of the short time interval between collective breakdown and the onset of the spread of pigmentation, one has to conclude that the extinguishing reaction only lasts for a short period. In this short time interval the activator has to drop to a level so low that a spontaneous re-activation is impossible even though the system rapidly returns to a bistable mode of activation.

Grüneberg (1976) postulated, on the basis of his investigation of shells of the *Clithon* family (see chapter 9), that two pigmentation systems exist, the leuco-pigment and the melano-pigment system. Both exclude each other and depend on each other. This schema is very similar to the mechanism derived here based on the dynamic properties of the system.

The longer and finer oblique lines in *Conus nobilis marchionatus* (Figure 7.2b) clearly underline the travelling wave mechanism. One of the most remarkable features is that the activation survives the collision of two waves. In contrast to the crossings discussed earlier (Figure 5.8, page 80), the two waves do not only penetrate each other. After collision, for an extended period no spread of activation takes place at all. The activation remains sharply localized. After this quiet period the spread of pigment deposition begins very slowly but, as indicated by the curvature of the lines, it becomes progressively faster at later stages until the next collision occurs. The resulting pattern is reminiscent of tilings or a pattern of staggered wine glasses. As in *C. marmoreus*, pigment deposition is switched off after some delay but, in contrast, this does not occur simultaneously in a large region. Instead, pigment termination starts more locally but with greater speed, even from the beginning. It catches up with the pigmentation wave but, since it is triggered by the latter, it cannot pass it.

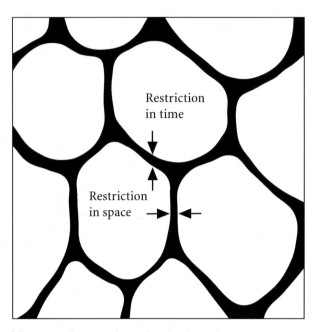

Figure 7.3. Pattern of "staggered wine glasses". The bent lines have an approximately constant thickness. Pigment production is alternately restricted either in time or in space.

It is very remarkable that the width of the pigmentation lines in this pattern remains approximately constant (Figure 7.3), although very different processes contribute to this constancy. Immediately after a collision, the width of the line perpendicular to the growing edge - the "stalk" of the wine glass pattern - is determined by a *spatial* restriction: the width in which activation survives. Later, after the wave speeds up, the width of the oblique lines is determined by a restriction in *time*: the duration of the pigment producing phase which becomes successively shorter and shorter. Both processes are, in principle, independent of each other. Later we will see the same phenomenon on other shells (for instance, Figure 7.9). This constancy rules out a large class of models in which survival is based on a gradual transition into longer activation phases (Figure 7.4). It also provides a strong argument for the assumption that what changes in complex patterns are parameters influencing the duration of the pigment production phase, i.e., the lifetime of one of the substances or rate of substrate production. In contrast, parameter changes that influence the absolute concentrations, such as changes in the rate of the autocatalysis, are presumably not involved.

To form the stalks some type of local signal must be present that enables an activation period at least ten times longer in a stalk region than in regions that are directly adjacent. The separate extinguishing reaction assumed here includes a possible mechanism. The two extinguishing waves may come to rest before they collide. A gap would remain in which the activation survives. The waves stop if the substrate of the extinguishing reaction is highly diffusible (Figure 7.2d, f) since the waves cannot enter the region depleted by a counter wave. New pigmentation

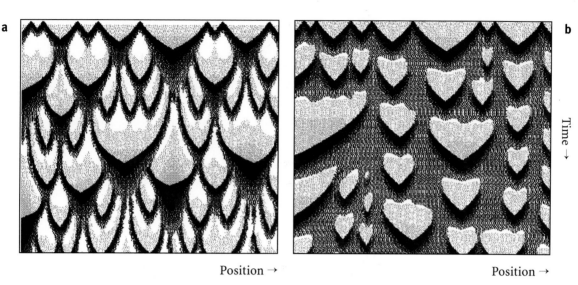

a

b

Time →

Position →   Position →

Figure 7.4. Two examples of patterns not found in nature. (a) Survival of collisions by the gradual transition to steady state activator production during long non-activated phases. The substrate (green) creates a feedback that effects the rate of its own production. During long non-activated periods substrate production reaches a level sufficient for a temporary steady state. (The rate of substrate production, not its concentration, is decisive in determining whether oscillations or a steady state will be formed, see Figure 3.3). At the end of these phases local maxima arise which begin moving toward non-activated regions. The gradual increase in line width and the resulting strong overshoot are arguments against such a mechanism. (b) Activation of pigment production reduces substrate production. After a certain period of pigment production, a transition into pulse-wise activation takes place and *vice versa*. Alternations between pigmented and non-pigmented phases occur but no tendency towards stalk formation exists. The white regions appear in columns since an overshoot during the onset of pigmentation increases the probability of another switch [S73a, S73b].

waves are formed after the refractory period of the neighbouring cells is over. Finer lines result if the substrate of the extinguishing reaction is produced not only by the pigment-producing cells but everywhere. With this condition the pigment reaction can trigger the extinguishing reaction more directly. Although the simulations reproduce essential features, we will learn that the mechanism must be modified for more complex shell patterns.

7.4 Resolving an old problem with the separate extinguishing reaction

An additional extinguishing reaction has been introduced to account for large scale breakdown. There are, however, much simpler patterns that indicate some sort of extinguishing reaction as well. Many shell patterns show a step-like rise and fall of pigmentation with a plateau at least as long as the intervals between the pigmentation periods. This is difficult to obtain with a single activator-antagonist system. If the duration of the pulse is elongated due to saturation, the pulse has a bell, not a step-shaped profile (see Figure 3.3 and 3.5). Step-like behaviour is,

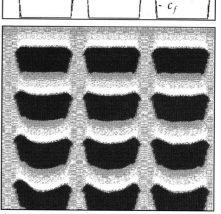

Position →   Position →

Figure 7.5. The possible role of the extinguishing mechanisms in simpler patterns. Switching off pigmentation using a separate extinguishing system allows constant pigmentation over a long time interval, in contrast to a system with an activator and antagonist alone. In the simulation of *Conus ebraeus* a bistable pigmentation system (black) is assumed. Pigmentation abruptly terminates due to the extinguishing reaction (red). This reaction is stable at some positions and suppresses pigmentation there permanently. In the remaining regions it is triggered regularly after a certain period of pigment production. The result is, so to speak, a reversed fishbone pattern (see Figure 4.10, page 64) [S75].

however, a straightforward outcome of a system in which bistability is switched off by a separate extinguishing reaction. An example is given in Figure 7.5. The accumulating substrate of the extinguishing reaction has no influence on pigmentation by itself. An abrupt change occurs only after a certain threshold level is exceeded and the extinguishing system is triggered. As discussed below, the bistability of the pigmentation reaction may be in itself the result of a composite process. The extinguishing feature, i.e., a hidden change over a long period culminating in a sudden overt change, may be used as a guide for designing models with more complex interactions.

7.5 The next step in complexity: the addition of a stable pattern

The patterns on *Conus episcopus* (Figure 7.6) and *Conus pennaceus* (Figure 7.9) display the same elements, white drops and fine oblique lines with "stalks" where two lines merge. However, the light brown background pigmentation is not without structure. Darker lines perpendicular to the growing edge are clearly visible. The enhancing pattern obviously has an influence on the initiation of the white drop pattern. Small white patches appear along the darker lines like pearls along a cord. Sensitivity to pigment termination depends on the position on the shell and changes over the course of time. The white drops may become larger and several may fuse resulting in relatively large non-pigmented regions. The subsequent

Equation 7.1: Pattern formation involving three systems: pigmentation, enhancing and extinguishing systems

The pigmentation system (an activator - substrate system):

$$\frac{\partial a}{\partial t} = s\, b\, a^{*2} - r_a a + D_a \frac{\partial^2 a}{\partial x^2} - c_e e a \tag{7.1.a}$$

$$\frac{\partial b}{\partial t} = b_b\, (1 + s_b c + c_b d) - s\, b\, a^{*2} - r_b b + D_b \frac{\partial^2 b}{\partial x^2} \tag{7.1.b}$$

The enhancing reaction (an activator - inhibitor system):

$$\frac{\partial c}{\partial t} = r_c\, a \left(\frac{c^2 + b_c}{s_d + d} \right) - r_c c + D_c \frac{\partial^2 c}{\partial x^2} \tag{7.1.c}$$

$$\frac{\partial d}{\partial t} = r_c\, a\, (c^2 + b_c) - r_d d + D_d \frac{\partial^2 d}{\partial x^2} + b_d \tag{7.1.d}$$

The extinguishing system (an activator - substrate system):

$$\frac{\partial e}{\partial t} = r_e\, f\, e^{*2} - r_e e + s_f a + D_e \frac{\partial^2 e}{\partial x^2} \tag{7.1.e}$$

$$\frac{\partial f}{\partial t} = b_f a + c_f(x) - r_e\, f\, e^{*2} - r_f f + D_f \frac{\partial^2 f}{\partial x^2} \tag{7.1.f}$$

with $\quad a^{*2} = \dfrac{a^2}{1 + s_a a^2} + b_a \quad$ and $\quad e^{*2} = \dfrac{e^2}{1 + s_e e^2} + b_e$

$-c_e e a$ describes enhanced a-activator destruction under the influence of the extinguishing system e.

$s_b c$ and $c_b d$ The increase of b-substrate production in the pigmentation system due to activator c or inhibitor d of the enhancing system.

$b_f a$ Substrate production in the extinguishing reaction caused by the pigment activator. This term plays a crucial role in patterns with large pigmented regions (Figure 7.2 c).

$c_f(x)$ Substrate production in the extinguishing system that is independent of the pigment activator. It plays a decisive role in patterns with fine pigment lines (Figure 7.2 d). It may depend on position (as indicated in Figure 7.6).

$s_f a$ The trigger for the extinguishing reaction caused by the pigment system. It is necessary when the substrate in the extinguishing system is produced independent of pigment reaction ($c_f > 0$, Figure 7.2d).

spread of pigmentation may result in two distinct patterns. Either narrow pigmented lines are formed or the striped background pigmentation is re-established.

With a single autocatalytic reaction one can either achieve full activation or a suppression. This is true even when two antagonists are involved (see chapter 5). Therefore, the three levels of pigmentation, very low in the white drops, higher in the background and even higher in the enhanced regions, require the superimposition of two systems. The existence of a stable pattern also resolves a problem remaining from the earlier model. Neither the bistable pigmentation reaction nor the extinguishing system are capable of initiating pattern formation. Simulations such as those shown in Figure 7.2 must be initiated from a set of specifically activated cells. Uniform activation would be wiped out by a single extinguishing reaction since the extinguishing waves come to rest only at a boundary between pigmented and unpigmented cells, or by the near-collision of two extinguishing waves. Both features are absent when activation is homogeneous. The missing ingredient is a spatial pattern whose local maxima act as nucleation centres for the extinguishing reaction. The fine lines in *Conus episcopus* and related patterns indicate that this feature exists.

The similarity between basic pattern elements and those discussed above, white drops and fine oblique lines (Figure 7.2), suggests that a third system should be added to the model which locally enhances pigment production. The complete system would contain the following components:

(i)   A bistable system that is responsible for the background pattern (light brown in Figure 7.6). It can spread slowly into non-pigmented areas.

(ii)   A stable pattern forming system (green in the simulations) that enhances pigment production causing darker lines perpendicular to the edge.

(iii) An extinguishing system (red) that accomplishes the sudden change from bistable to pulse-wise activation, for instance, by more rapidly destroying the activator in the pigmentation system.

Some features of natural patterns restrict possible interactions. The darker lines are narrower than the spaces that separate them, suggesting that the stable pattern is active where the dark lines occur and that this causes an enhancement of pigmentation. The reverse possibility, that the pattern is in an active state in the regions of light pigmentation and reduces pigment production, can be excluded. Since the dark lines are never observed to shift to neighbouring positions, the enhancement system is more likely realized by an activator-inhibitor rather than an activator-substrate system (see Figure 2.7 for the different properties of these two systems). The white patches appear preferentially on the dark lines, indicating that the stable system enhances the probability of initiating pigment termination. This can be integrated into the model outlined so far in a straightforward manner. The stable system is assumed to increase substrate production and therewith the activator concentration of the pigmentation reaction. This leads to darker lines. In turn, it was assumed that the substrate of the extinguishing reaction was produced

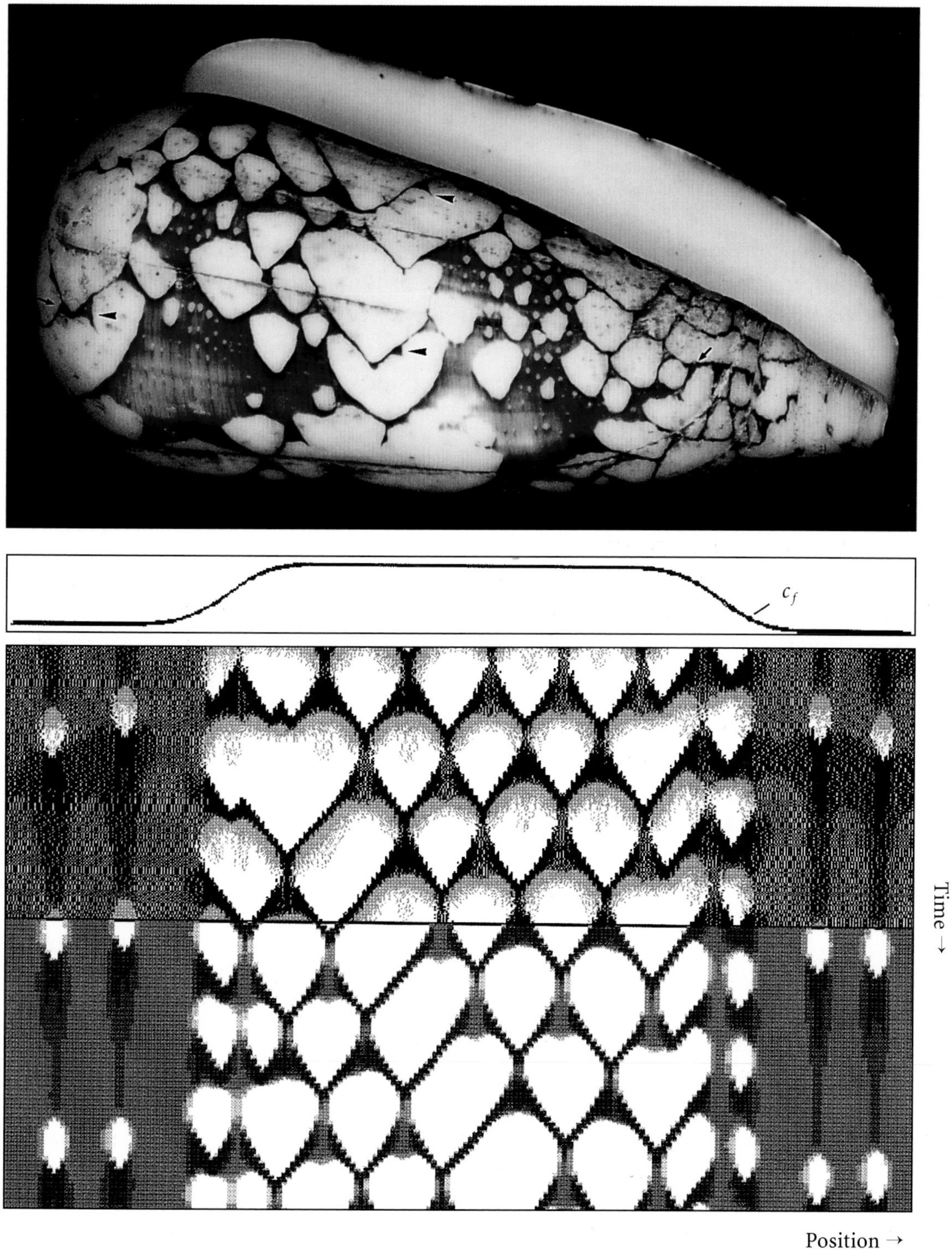

Figure 7.6. *Conus episcopus.* Large pigmented areas display darker lines with small white drops. The fine oblique lines form branches (arrow) or attempt to branch (arrowheads). In the simulation three interacting pattern forming systems are assumed: the pigmentation (black), the enhancing (green), and the extinguishing system (red). The upper part shows all activators, the lower only the expected pigmentation. The simulations are calculated with equation 7.1, page 111. The graph above the simulation indicates the substrate production for the extinguishing reaction independent of pigmentation, $c_f(x)$. The higher rate in the centre leads to larger unpigmented regions. [S76].

proportionally to the activator concentration of the pigment producing system. Therefore, the extinguishing reaction is fired more frequently along the lines of highest pigmentation.

7.6 Mediating branch formation with a temporary stable pattern

The extended model correctly describes some basic features. For instance, that white drops along lines of enhanced pigment production are much wider than the lines; or that two colliding waves frequently survive, causing a Y-shaped pattern element. In contrast, the formation of branches and attempts to branch are not reproduced. In *Conus episcopus* (Figure 7.6) these elements are not very obvious, but in *Conus pennaceus* BORN (Figure 7.7) branches are the dominating pattern element. Thus, a model with a claim to generality must account for branches as well. We shall see that the modifications required for branch formation will allow us to describe many other patterns as well.

In the model outlined thus far, the formation of stalks results from the non-collision of two approaching waves of the extinguishing reaction. Correspondingly, branch formation would require a sudden decrease of an extinguishing wave. But what would signal such a decrease when no counter wave is present? This is only one of the problems with the simulation of stalk formation by non-colliding waves. In the simulation of *Conus nobilis marchionatus* the stalks are very thin (Figure 7.2d), in contrast to the real pattern. It is not a question of parameters. If the extinguishing reaction is too weak, a direct spontaneous re-activation can take place; if it is too high, no stalk formation is possible. Moreover, in many shells the level of pigmentation increases shortly before branch or stalk formation - again, a feature incompatible with the assumption that stalk formation results from an arrest of extinguishing waves.

The enhancing reaction introduced above to account for darker lines resolves these problems. Imagine that the pigmentation reaction generates travelling waves. The enhancing reaction can be triggered along these waves with a certain spacing due to the long range inhibition. This increases the local substrate supply of the pigmentation reaction and elongates the pigment producing period. If this elongation lasts longer than the refractory period of neighbouring cells, branch formation is initiated (Figures 7.7 and 7.8a, b). Otherwise it causes an attempt to branch, resulting in an irregularity on the lower edge of a line. Stalk formation would require triggering of the enhancing reaction around the point of collision.

7.7 Generating a steady state with two mutually stabilizing reactions

In the simulation of the white drop pattern (Figure 7.2), steady state pigment production was an essential ingredient. The formation of branches suggested that

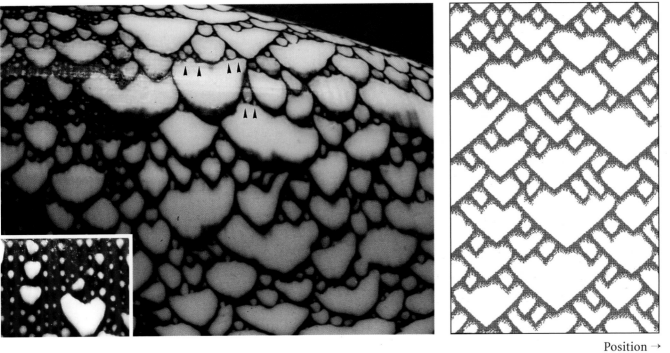

Position →

Time →

Figure 7.7. Branches as dominating pattern elements: *Conus pennaceus* BORN. Other elements indicate a close relationship to the pattern shown in Figure 7.6. Small white patches are arranged along darker lines (insert) and larger white regions result from abrupt pigment termination over an extended region (arrowheads). Simulated with the same model as Figure 7.8b; [S77].

this steady state is the result of special enhancing reaction rather than an intrinsic property of the pigment reaction on its own. The assumption that both patterns result from a common mechanism suggests a generalization: steady state pigment production in complex shell patterns results from the interaction of two patterns that mutually depend on each other. One acts as a precondition, the other varies the time constants of the first, moving towards a steady state. What is most interesting in this explanation is that the same mechanism accounts for very different patterns depending on the character of the two systems involved. These pattern elements are basic components in complex shell patterns. The following cases can be distinguished (Figure 7.8).

(a) Both the pigmentation and the enhancing patterns oscillate. As described earlier, branches are formed.

(b) The pigmentation pattern oscillates but the enhancing pattern is stable over time: separate lines parallel to the direction of growth are formed from which travelling waves spread in both directions. A fishbone-like pattern results (see also Figure 4.10).

(c) The pigmentation and the enhancing patterns are in a steady state: dark lines parallel to the direction of growth emerge on a pigmented background.

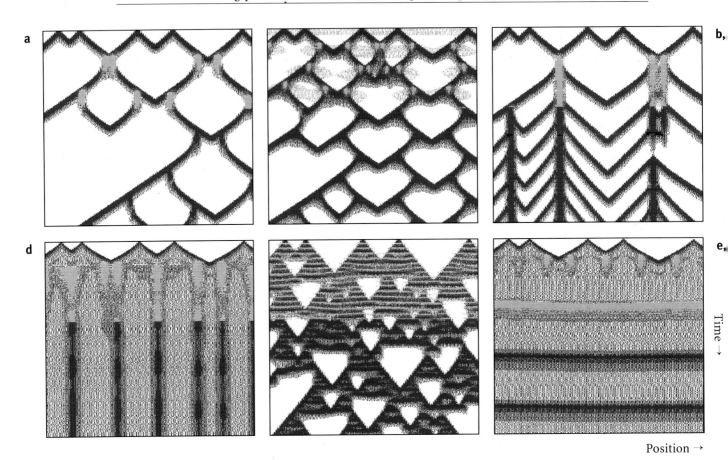

Figure 7.8. Patterns generated by two systems that mutually depend on each other. (a) The pigmentation reaction generates travelling waves on its own. The enhancing reaction (green in the upper parts) requires the pigmentation reaction and is spaced by a diffusible inhibitor. If triggered, it causes a local increase in substrate supply and thus an elongation of the activation. Branch formation is possible. (b) The inhibitor (red) of the enhancing reaction rather than the activator causes increased substrate production. The shapes of branches are reproduced better. (c) The enhancing reaction is stable in time. Lines parallel to the direction of growth are formed from which travelling waves spread in both directions. (d) The background and the enhancing pattern are stable. Dark lines appear on a lighter background. (e) A rapid, nearly synchronously oscillating enhancing pattern leads to a near steady state over long time intervals. Large scale breakdowns cause unpigmented triangles. (f) The background system is in a steady state; the enhancing system oscillates. Darker lines of pigmentation parallel to the growing edge result. Calculated with equation 7.1 a-d; [S78a-S78f].

(d) Both the pigmentation and enhancing systems are on the border between a steady state and oscillations. Large but irregularly sized pigmented regions and finer oblique lines may coexist in the same pattern. This type of pattern will be discussed again in the next chapter (Figure 8.11).

(e) The pigmentation pattern is in a steady state but the enhancing pattern oscillates: lines parallel to the growing edge are formed. This type of pattern will also be discussed again later (Figure 9.8)

Several interactions between the enhancing and pigmentation reactions are conceivable. Either component of the enhancing system, the activator or the inhibitor, can mediate the increase in the substrate production of the pigmentation reaction. Figures 7.8a,b show the corresponding simulations. The branches have slightly different shapes. If the inhibitor is involved, the enhancement is less localized due to the inhibitor's longer range and lifetime. The initiation point for a branch has a more triangular shape. In contrast, the more localized activator produces a sharper line from the beginning. A comparison with natural patterns indicates that the inhibitor solution corresponds more closely. Later we shall encounter more evidence for the involvement of antagonists in the interference between two reactions (Figure 8.7).

7.8 Intimate coupling of enhancing and extinguishing reactions

Although the use of an enhancing reaction for local survival describes branch formation very well, the situation is unsatisfactory because it leads to two different models for closely related patterns. On the one hand, for the white drop pattern we are assuming that the pigmentation reaction remains in a steady state and the extinguishing reaction creates the white drops (Figure 7.2). On the other hand, for branches we are assuming that pigmentation produces travelling waves on its own and a special enhancing reaction is required for the temporary transition to a steady state. Do we need both modifying reactions, enhancement and extinguishing?

The shell of *Conus pennaceus* (Figure 7.9) provides a clear answer. There, dark lines and white drops regularly alternate in some regions. This cannot be the result of a periodic enhancing reaction alone since this would only cause an alternation between darker and normal background pigmentation but not white drops (see Figure 7.8d, f). Hints of an extinguishing reaction also exist in patterns that are ruled mainly by branching, i.e. in patterns where enhancement is the prevalent element. In Figure 7.7, abrupt large scale pigment termination takes place after the formation of many branches within a short time interval (arrows). Therefore, both elements are necessary.

If the extinguishing reaction is accomplished by a separate non-linear mechanism, it must be an all or nothing event. However, circumstantial evidence exists in favour of a more continuous transition. In *Conus pennaceus* there are several instances of ladder-like connections between two diverging lines (Figure 7.9b, d). This suggests that the steady state causing the stalks smoothly changes to oscillating conditions. Such a ladder-like pattern is never observed in simulations in which stalks are terminated by a non-linear extinguishing reaction (Figure 7.2d).

Since the white drops preferentially appear along the darker lines, the enhancement and extinguishing reactions must be coupled. The situation appears to be paradoxical. A modifying reaction must be involved that has a stabilizing *and* a

a

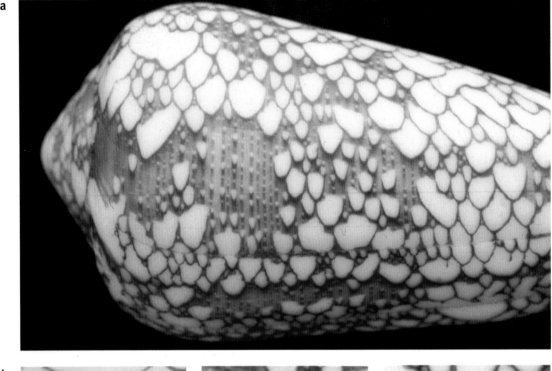

b 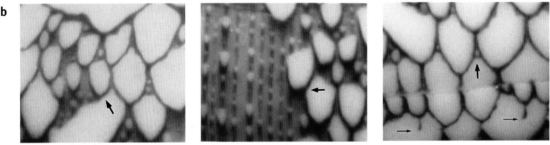 c, d

Figure 7.9. *Conus pennaceus.* (a) Small drop-like patches and a pattern of "staggered wine glasses" on the same shells. In regions with coherent background pigmentation darker lines appear that alternate with white drops. The white drops appear to have a shadow. The darker pigmentation results from an overshoot after the onset of the pigment reaction. (b-d) Details that restrict possible models. (b) On one side of a white region a fine line appears, on the other side background pigmentation. No stalk formation takes place. Obviously, the collision of two pigmentation waves is insufficient for stalk formation on its own. (c) A fine line ends in a "shadow". Activation in the overshoot region lasts much longer than in the fine line region. This supports the view that either the fine lines result from an active shortening of activator half life or the background pigmentation results from its active elongation. (d and b) Switching off pigment production in the stalk region need not be complete. It may occur over several oscillations. On the shell this causes one or more ladder-like connections between the two diverging lines (large arrow). This is an argument against an enforced extinguishing reaction and in favour of a smooth change from a steady state to an oscillating regime. After collision the narrow perpendicular lines occasionally end blindly (small arrows). Obviously, the refractory period of the neighbouring cells can be so long that a re-infection is not possible before pigment production ceases. This phenomenon is also visible in *Conus episcopus* (see Figure 7.6).

destabilizing influence. The nature of this coupling is the central problem in modelling complex patterns and will be a constant theme throughout the remainder of this book. Again, the problem is that many possible implementations are conceivable. One of them has already been discussed when simulating *Conus episcopus* (Figure 7.6): the enhancing reaction produces more pigment and this pigment accelerates the trigger of the extinguishing reaction. In this example, both modifying reactions were complete pattern forming systems and only loosely coupled. Are other reactions conceivable in which both features result more naturally from a single reaction?

A common feature of cell differentiation in the development of higher organisms is that one particular differentiation is followed by another. In shell patterns, however, several observations argue against a periodic alternation between enhancing and extinguishing phases. In Figure 7.9 the white drops are definitely wider than the lines of enhanced pigmentation. Therefore, the enhancing reaction cannot be a precondition for breakdown. Similarly, the irregular alternation between dark and white patches on a pigmented background (see Figure 1.9, page 13) excludes the possibility of both processes resulting from different phases of a single cyclic reaction.

It may appear reasonable to assume an interference from a fast enhancement reaction and, on a larger time scale, destabilization after the sufficient accumulation of a "poison". This, however, will not work. All destabilizing influences (shortening activator half-life, elongating inhibitor half-life, or reducing substrate production) also cause a decrease in activator concentration and would thus lead to a lower rate of "poison" production. Rather than alternating between enhanced pigmentation and white drops the consequence would be a permanently reduced steady state. As mentioned earlier, it is crucial for the extinguishing reaction that hidden changes occur without any visible alterations until a certain threshold level is obtained.

7.9 Extinguishing that results from a depletion of resources due to an enhancing reaction

One key to the above problem may be found in the patterns of *Conus pennaceus* (Figure 7.9) and *Conus auratus* (Figure 7.10). The latter shows long enhanced lines uninterrupted by white drops. These phases of coherent background alternate with phases in which the background pigmentation disappears and only narrow oblique lines with crossings remain. White drops are only formed in the zones of transition between the two phases. In other words, white drops occur only when the system is near the border between a steady state and a pulsating mode of pigment production. This suggests that the enhancing reaction is accompanied by a depletion of resources, causing an earlier breakdown. The situation may be compared with doping in sport. Doping releases the available resources of the body, enabling temporarily high output of power. This, however, can cause a breakdown in the body when resources become exhausted. Without doping, a continuous output of power is possible over a longer period of time although at a lower level. Exhaustion comes into effect only if the system is at the border where continuous activation is just possible. In contrast, if the ability to replenish resources is sufficiently high, no enhancement-induced breakdown will take place.

Several implementations of this idea are conceivable. For instance, a pool may exist from which the substrate for the autocatalytic reaction is obtained. The enhancing reaction would allow an increased utilization of this pool with a corresponding depletion. The dynamics of such a system, however, is difficult to understand. For instance, it seems reasonable to assume that in a normal activator-substrate system (Equation 2.4), the rate of substrate removal in a cell depends on whether the cell is activated or not. However, this is not the case. This is most easily understood by realizing that, by definition, the concentration change is zero in the steady state. Since the production rate b_b is the same in all cells, the sum of all processes that lead to substrate removal (consumption in the reaction, loss by diffusion, and independent decay) must balance constant production. It must also be the same in all cells and independent of local activation. Therefore, if a pool exists from which substrate is produced, utilization of the pool would be independent of local activation. If, however, a lower substrate concentration leads to an increased transfer from the pool to substrate production, a preferential depletion will occur.

The system satisfies the requirement that accumulating changes must be hidden. As long as sufficient precursors are available, the system works normally. The situation may be compared with a car whose output of power is independent of how much fuel is in its tank. Of course, the situation changes suddenly when the tank is empty. The substrate pool differs from the car analogy though, because there is a constant influx into the pool even when it is almost empty. But this supply may be insufficient to maintain the steady state. Figure 7.10 has been calculated using this scheme. It can be seen that the first white patches appear in those regions that were previously exposed to the enhancing reaction. However, the correct simulation of white drops along dark lines using a single modifying reaction is not yet achieved.

$\longrightarrow$

Figure 7.10. *Conus auratus.* Periods of coherent background pigmentation alternate with meshwork-like structures over long time intervals. Dark lines of enhanced pigmentation are clearly visible on the light brown background. Small white patches are formed only in zones of transition. Model: a pool of precursor molecules for the production of the substrate is assumed. If the influx into this pool is sufficiently high, pigmentation reaction is in the steady state, otherwise travelling waves are formed. The enhancing reaction (green in the upper part) is responsible for the stripes of increased pigmentation in the steady state as well as for branching in the travelling wave mode. Regions exposed to the enhancing reaction are the first to terminate pigment production. To simulate the transition from a steady state to oscillatory behaviour, the influx into pool b_e has been changed from 0.06 to 0.035 and back to 0.06. The time of each change is indicated by an arrow. In reality, this is certainly accomplished by an independent oscillation; [S710].

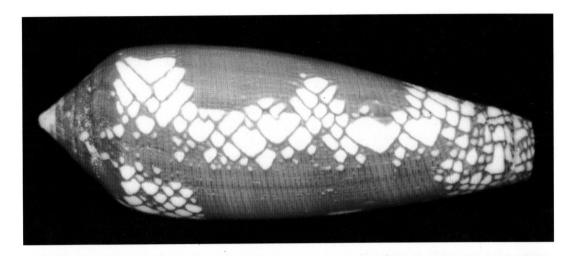

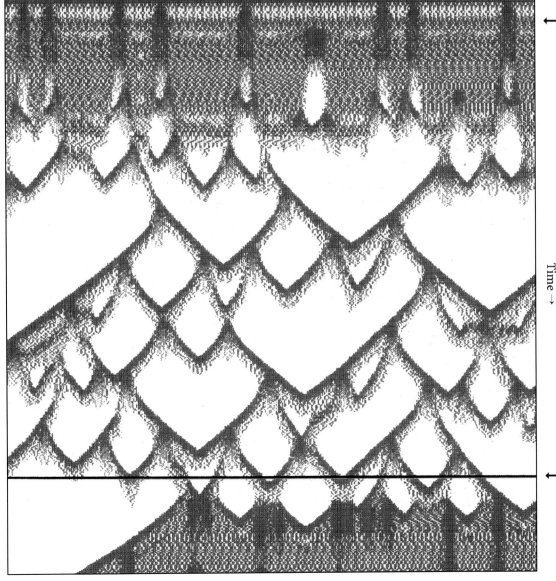

Time →

Position →

7.10 Related patterns reveal open problems

The patterns on related shells demonstrate a wide range of variations. The shell of *Conus ammiralis* (Figure 7.11a) shows in some regions the usual patterns of white drops and of dark stripes on a pigmented background. In other regions, however, the coherent background pigmentation is replaced by a meshwork of fine lines. It is quite remarkable that the size of the large white drops, their frequency, and the spread of the pigmentation are not affected by the transition from one region to another. This was not be expected from the models outlined above. It was assumed that enhanced pigmentation preferentially triggers the extinguishing reaction. This accounts for chains of white drops along the dark lines (Figure 7.6). But in *Conus ammiralis*, pigment termination seems to work normally even if much less pigment is produced. A similar problem exists with another pattern discussed earlier (Figure 7.9). Increased pigmentation occurs in the "shadow" of the white drops resulting from an overshoot of the pigmentation reaction. Why does this not lead to an immediate extinguishing reaction?

Figure 7.11b shows a shell of a closely related snail, *Conus ammiralis archithalassus*. In addition to the normal lines of enhanced pigment production perpendicular to the growing edge, dark lines parallel to the edge are also visible. The large white drops are more irregularly arranged but, especially the finer ones, appear preferentially at a crossing of the two types of enhanced lines. This underlines the intimate coupling between enhancement and termination of pigment production. The fine meshwork suggests that steady state pigment production is in itself a composite process and is based on an enhancement. If this enhancement is insufficient to maintain an overall steady state, its oscillation causes a rapid sequence of branching. Given this interpretation, the thickness of the fine lines would be a measure of activator half-life without enhancement.

The rounding of white drops at their lower tip is a special problem (compare, for instance, the collision of waves in Figure 6.1 and Figure 7.9). These roundings indicate that two approaching waves are speeding up. Two mechanisms can be envisioned: (a) The cells become more excitable the longer they have not been activated. The excitability of the cells, and thus the speed of the wave, is correlated to the length of the non-activated period. (b) Alternatively, two approaching waves are attracted to each other; the speed would correlate with the remaining distance between the two waves. From the pattern one can infer that the rounding is roughly the same in large and small white drops, giving more support to the second possibility.

7.11 Apparently different patterns can be simulated by closely related models

Several shells that overtly look very different can be described with basically the same mechanism. The shell of *Cymbiolacca wisemani* (Figure 7.12), for instance,

a 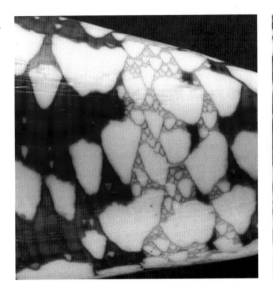  b

Figure 7.11. Pattern variations in related species. (a) *Conus ammiralis*. Its most remarkable feature is the transition from regions of pigmented background with dark stripes to light narrow meshwork-like pigmentation. The size of the white drops and the spread of pigmentation remains the same in both regions. (b) Shell of a related species, *Conus ammiralis archithalassus*. Darker pigmentation lines parallel and perpendicular to the growing edge are visible. The fine meshwork pattern is restricted to narrow stripes. They are almost free of white drops.

shows unpigmented patches. Some of them have dark dots at the top. No fine lines or stalks are present. However, a closer inspection reveals that the same elements are involved here as have been discussed earlier. The white patches appear due to abrupt large-scale termination and slow re-pigmentation. The dark dots at the beginning of the white patches underline the postulated intimate coupling between enhancing and extinguishing reactions. Again, the latter extends further. Both events, enhancement and extinction, follow rapidly on each other. It is difficult to decide whether a local increase of pigmentation enhances the probability of a large scale extinguishing reaction or whether it is a part of the extinguishing reaction itself. The pigment production must be in the ON state for some time before the enhancing reaction, and thus the destabilization, can be triggered. It is quite remarkable that large scale pigment termination can occur without an enhancing reaction beforehand, especially in regions of high background pigmentation. Again, the enhancing reaction is not a prerequisite for pigment termination. The main difference for the patterns discussed above is that the enhancing pattern obviously has a destabilizing effect only on background pigmentation.

The shell of *Marginella limbata* (Figure 7.13) shows common travelling waves. In addition, very fine dots are visible that are strictly arranged in rows. Patches of dark pigmentation at the growing edge mark some of these rows. Three pattern forming systems must contribute: the first forms an invisible but permanently stable pattern that acts as precondition for the fine dots. The second forms the waves and the third is triggered only by the simultaneous activation of the first two

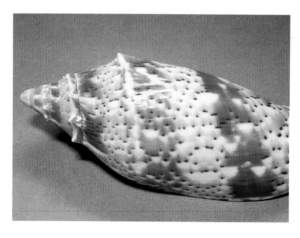
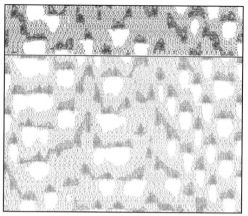

Time →

Position →

Figure 7.12. Unpigmented patches with an i-dot, the shell of *Cymbiolacca wisemani*. The same pattern elements are involved: local enhancement and concurrent large scale breakdown. An enhancing pattern that leads to staggered dots (see Figure 5.10) is assumed in this simulation. Increased pigmentation accelerates the trigger for the extinguishing reaction [S712].

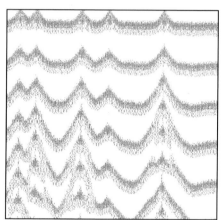

Time →

Position →

Figure 7.13. Travelling waves and rows of fine dots: *Marginella limbata*. Model: The coincidental activation of both a stable and a wave-generating system can trigger a third system, an enhancing reaction, for a very short time. Small dots are formed in rows. Sometimes two such dots are formed within the passage of a single wave (arrows). This pattern indicates that a stable system may be a prerequisite for the enhancing reaction, but since it switches off so rapidly it seems that activation of the enhancing reaction has its own dynamics [S713].

systems, i.e., if a wave passes an active region in the stable pattern. The third system must have very short time constants, since the rise and decline of pigmentation occur very rapidly. The short time constants and the required presence of the two other systems lead to the formation of dots. A very low concentration of the wave system is sufficient for the trigger. Therefore, the dots appear to arise shortly before the wave passes. Sometimes two such bursts fit into one stroke of the wave. This can be the origin of a wave running in the opposite direction. The most important feature in the present context is that switching off the enhancing reaction is an independent and very rapid process. This shell supports the view that the superimposition of waves on a stable pattern leads to pigment enhancement. It is special in that the enhancing reaction has two clearly separable components, one with a long and another with a very short time constant. The patterns of spine formation in Figure 4.1 certainly have a similar basis.

7.12 Conclusion

The complexity of many shell patterns results from the superimposition of the pigmentation system with two types of modifying patterns. An enhancing pattern increases pigment deposition and frequently shifts the pigmentation system from an oscillating in a temporary steady state. Branches, the formation of darker lines on a pigmented background, and the formation of patches with increased pigmentation are traces of this reaction. A second modifying influence has the opposite effect causing an abrupt and often large scale termination of pigmentation.

The number of possible interactions between these three systems is extremely large. Deducing the most likely form of interaction from the details of a shell pattern is a very difficult task. The interactions given here should be regarded only as examples.

The postulated systems have different properties - a feature that makes further simplifications difficult. The pigment system requires a slowly diffusing activator and a non-diffusible antagonist in order for the pigmentation reaction to spread. The components of the extinguishing reaction must spread rapidly to enable the large scale break down. The enhancing reaction must have a rapidly spreading antagonist with a short time constant in order to generate at least a temporary stable pattern. Since this pattern may change after long inactive periods, presumably it depends on the background system. It is not yet clear whether the extinguishing reaction is based on an independent system or results as an implicit consequence of the enhancing system. Arguments for both versions have been discussed. In any case it requires some types of large scale action.

On the other hand, both modifying influences are not always required in order to generate interesting patterns. Without the enhancing reaction simple white drop patterns are obtained (Figure 7.2). With the enhancing but without the extinguishing reaction, the huge class of patterns discussed in chapter 4, the net-like patterns in Figure 7.8 and the pattern characterized by large pigmented areas and finer oblique lines in irregular arrangement (see Figure 8.11) are obtained.

The generation of higher organisms requires the superimposition of many pattern reactions. Although biological systems handle this complexity without any problems it makes theoretical analysis very difficult and the result, even if correct, may appear inelegant. Inelegance, however, is an aesthetic problem only for theorists.

Chapter 8

Triangles

Several molluscs display triangles as their basic pattern element. The triangles may be connected to each other to form oblique lines with a triangular substructure. If both corners of the lower edge give rise to new triangles, the white regions in between also have a triangular shape although with opposite orientation. The triangles may cover different portions of the shell. If they are densely packed, it appears as if white triangles are arranged on a black background. The triangles can also be of very different sizes. On some shells they are a prominent pattern element, on others they appear more as a roughness in the oblique lines but are clearly visible on closer inspection. The triangles themselves may have a fine structure of lines parallel to the growing edge or they may resolve into bundles of lines parallel to the direction of growth. On some shells an almost continuous transition from triangle to branch formation can be recognized. The occurrence of triangles on very different molluscs, on bivalved mussels and on snails, indicates that the possibility of forming triangles is a basic feature of shell patterning. Figure 8.1 gives some examples. In this chapter, an attempt will be made to find a unified explanation for this diversity. I will begin with the basic features and how they can be modelled within the framework of the theory. Discrepancies with natural patterns will be used as guides to develop more complex models.

As mentioned in chapter 7, formation of triangles requires a bistable system. From a single activated point, activation spreads in both directions and the cells remain in an activated state. While in the patterns discussed in the previous chapter termination of pigmentation spreads more rapidly than the onset of pigmentation (Figure 7.2), the sharply straight lower edge of triangles indicates that termination occurs strictly simultaneously. This excludes the possibility of a signal being initiated at a particular position spreading by diffusion. Therefore, it is assumed that the signal for pigment termination in triangle formation does not result from a metabolic product of the pigment producing system but from an independent oscillating system. The substances responsible for the extinguishing reaction can be distributed within the animal in hormone-like fashion. In this way concentrations of the oscillating system are constant along the growing edge of the shell. More

Figure 8.1. Different shells bearing triangles as their basic pattern element

Figure 8.2. Simple model of triangle formation. (a) Details of *Lioconcha castrensis*. (b-d) A bistable reaction is assumed. Activation (black) can spread in both directions due to activator diffusion. An oscillation (red) is superimposed that blocks pigment reaction, either (b) by suppressing the necessary substrate production, or (c, d) by increasing activator removal. For a short time the activation is shifted from a bistable mode to a mode with short pulses. Activation can survive at the margins of the resulting triangle since here sufficient substrate is still available. Both models are parameter sensitive. An increase of 10% in the rate of activator removal by the extinguishing reaction leads to a dramatic change in the pattern (c *versus* d). A similar pattern alteration causing bilaterally connected triangles to separate into chains of triangles is also visible on the shell (a) [S82a, S82b, S82c].

or less synchronous oscillations as shown in Figure 3.4 (page 45) are conceivable as well.

The most surprising feature of connected triangles is the almost immediate spread of pigmentation after a collective breakdown. Usually there is no lag between the breakdown and the spread of pigmentation into the same region again. Thus, the signal for a transition from the steady state to the pulsewise mode of pigment formation must be very short. The system returns to its normal condition of bistable activation very rapidly. In connected triangles, one side should be in the shadow of a previous activation while at the other side, pigmentation spreads into a region that was not activated for a long period. Nevertheless, the shape of most connected triangles is symmetric, suggesting that the spread of pigmentation is independent of the time that the cells were not activated. In other words, the system does not become more excitable as time progresses from the last activation. A similar feature was mentioned in chapter 6 in connection with the branching of lines. In that case, backward waves proceeded immediately after their birth with the same speed as the original wave, although they were in the shadow of the primary wave (Figure 6.1). The inherent similarities between triangle and branch formation will be discussed again later.

Figure 8.2 shows simulations of triangle formation using simple models. An oscillating system (red) shifts a bistable system from active to inactive state for a short period. This transition can be accomplished either by a reduction of substrate production (Figure 8.2b) or by additional removal of the activator. The first mechanism is less convenient for producing symmetric triangles since it requires time for the recovery of substrate concentration. If the extinguishing reaction is short and of moderate strength it may require several pulses before the activator no longer recovers instantaneously. The result is a pattern similar to the famous Sierpinsky triangles with their fractal geometry. It closely fits the pattern on the shell of *Cymbiola innexa* (Figure 8.3).

8.1 The crossing solution through the backdoor

The models developed so far are parameter-sensitive. A small increase of the efficiency in the extinguishing reaction leads to a considerable change in the resulting pattern. Nevertheless, this may correspond to the actual situation since a single shell can exhibit a variety of analogous patterns (Figure 8.2). However, several distinct features of natural patterns are not reproduced by minimal models. Although pigmentation spreads with the same speed in both directions, activation often survives only on the outer side of the triangles. There must be a memory feature that does not affect the spread and comes into play only at the time of collective breakdown. A region recently activated would remain more prone to the extinguishing reaction. Figure 8.4 illustrates this situation.

Position →

Time →

Figure 8.3. Sierpinsky triangles: the pattern on *Cymbiola innexa* REEVE. The signal for the transition into pulse-like activation (red) may be so short that activation (brown) reappears immediately. Occasionally however, this may fail. This is relatively frequent in small areas, forming many small triangles. It may also occur, although more rarely, over a larger area leading to large triangles. The pattern is similar to the well-known Sierpinsky triangle with its fractal geometry [S83].

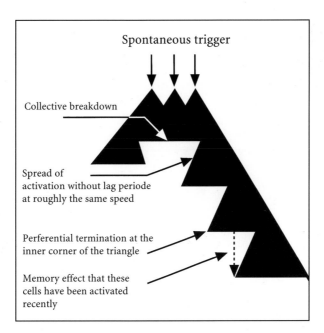

Spontaneous trigger

Collective breakdown

Spread of
activation without lag periode
at roughly the same speed

Perferential termination at the
inner corner of the triangle

Memory effect that these
cells have been activated
recently

Figure 8.4. The problem of unilaterally connected triangles. Although pigmentation spreads with the same speed in both directions after a breakdown, termination occurs preferentially on the side that produced pigment earlier.

Another problem is the spontaneous initiation of new chains of connected triangles. This is impossible using the minimal model described above. The symmetric shape of the triangles requires almost instantaneous recovery of the system and thus very short time constants. In contrast, a spontaneous trigger occurs only after a long time interval. Both the unilateral survival mentioned above and the delayed spontaneous trigger indicate that a second antagonist, for instance an inhibitor with a long time constant, is involved. This inhibition is higher in cells that have recently been activated, lowering the chances that the pigmentation system will survive the extinguishing reaction at the inner margin. A spontaneous trigger is possible after the decay of this second inhibitor. Figure 8.5 shows examples of different shells of the *Lioconcha* family and corresponding simulations.

A system with two different time constants was already employed for the formation of crossings (see Figure 5.8, page 80). It was mentioned that the oblique lines in these crossings show a fine substructure of connected triangles. The assumption of a second antagonist for these systems was based on the survival of activation during collisions. Now, the triangles described here use the same model for a different reason. The first antagonist is required for fast recovery after collective breakdown. The second one is responsible for the memory feature that leads to preferential termination of pigmentation on the inner side of the chain of triangles. Later we shall see that a third inhibition must also exist.

a

b

Time →

Position →

Figure 8.5. Chains of connected triangles. (a, b) Pattern on different species of *Lioconcha*, (c) *Sunetta meroe*. (d-f) Simulations using Equation 5.1 plus an oscillating system. The activator of the oscillating system (red) shortens the activator half-life of the pigmentation reaction for a short time interval. Depending on the parameter values used, the triangles may be either the dominating pattern element or more a fine structure. Occasional branching, crossing and annihilation after a collision are reproduced. In (f), the inhibitor of the oscillating system increases the substrate production of the pigment system (see Figure 8.7); [S85a, S85b, S85c].

8.2 Triangle *versus* branch formation

Although the occasional branching and the crossing of the chained triangles is correctly reproduced, finer details indicate that even this model is too simple. The triangles in the simulations are too regular when compared with natural patterns. Pigment termination is sometimes incomplete causing a fuzzy lower edge to the triangles. The time interval between termination may also vary.

Other shells of the *Lioconcha* family and related shells show further interesting modifications (Figure 8.6). In parts of *L. castrensis* (8.6a) that were formed at younger stages, triangles and branches visibly coexist. In *L. ornata* (8.6b) branching is the dominating structure. In *L. hieroglyphica* (8.6c), branching lines frequently terminate without collision or form triangles that have an internal structure of fine parallel lines (8.6d). The lines may become thicker and smeared (8.6e). In *L. lorenziana* (8.6f), pigmentation fades away rather than forming branches. In the remaining sections of this chapter, possible origins of this diversity will be discussed.

The simultaneous formation of branches and triangles on the same shell (Figure 8.6a) is very remarkable since, as explained above in detail, the two patterns require opposite sets of parameters. Branches are formed by a system of travelling waves in which a short temporary transition into a steady state takes place. The transition enables the survival of activation until the refractory period of the neighbouring cells is over. In contrast, triangle formation needs a pigmentation system in a steady state with short temporary transitions into pulse-like activations.

These conflicting requirements can be resolved under the assumption that both modifications, extinguishing and enhancement, occur in rapid succession. So far, we have assumed that the activator of the modifying system shortens the lifetime of the pigment activator, causing a transition from a steady state to pulse-like activation. The branching phenomenon suggests that the antagonist of the modifying reaction, for instance an inhibitor, plays the opposite role in this modification. While the short pulse of the activator acts as an extinguishing signal to form the lower border of the triangle, the longer lasting inhibitor that immediately follows brings the pigment system towards a steady state. Depending on whether the unperturbed pigmentation system forms travelling waves or tends to bistability, and on how strong the extinguishing and/or enhancing influences are, either triangles or branches will be formed. Simulations of both modes are shown in Figure 8.7. The upper figures provide a close-up of how the system works showing the pigmentation system (brown), the extinguishing activator (red) and the enhancing inhibitor (green) of the modifying reaction. The lower two panels provide simulations using the same parameter values over a larger field of cells to demonstrate the overall pattern. With this modification, triangle formation becomes even more similar to the extinguishing and branching patterns discussed in chapter 7 (see, for instance, Figure 7.8, page 116).

In order to maintain a bistable steady state, high substrate production and a high decay rate independent of the autocatalytic process are required (see Figure 3.3). In other words, much of the substrate produced would not contribute to pattern formation. The branching phenomenon suggests a more economic method of substrate production. It would increase only temporarily when required, i.e., immediately after an enforced breakdown.

8.3 The involvement of three inhibitory reactions

The shell of *L. hieroglyphica* (Figure 8.6c) shows oblique branching lines that terminate without any collision with counter-waves. This phenomenon is also visible

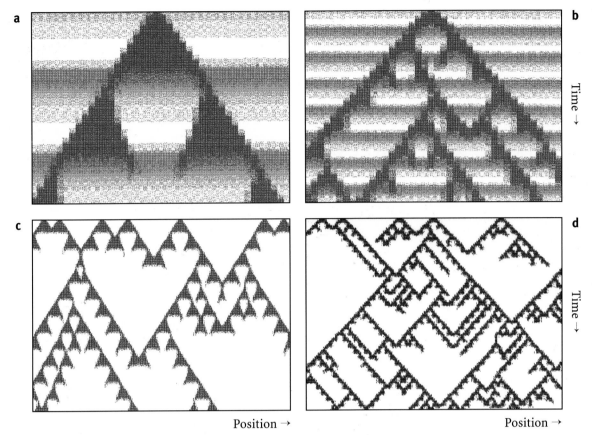

Time →

Time →

Position →   Position →

Figure 8.7. Triangle *versus* branch formation. Triangle formation requires an abrupt termination of pigment production, branch formation requires its enhancement. Both requirements can be met if the activator (red) and the inhibitor (green) of the modifying system have antagonistic influences on the pigmentation system. (a, c) The extinguishing effect of the activator dominates: triangles result. (b, d) The enhancing effect of the inhibitor dominates: branches result [S87a, S87b]

in a milder form in Figures 8.6a and 8.6f. Termination of this type is not announced by a retardation in the spread of pigmentation. This suggests that activation is near the lowest level at which the chain of triggering events can be maintained. Any additional lowering causes termination even in cells that have not been activated for long time. What can cause wave termination? Termination occurs preferentially if two waves approach each other. Either one or both waves terminate. In the latter case, a non-activated gap remains. This suggests that a rapidly spreading inhibition emanates from the activated cells. Sensing this inhibition a counterwave may terminate. If the time constant of this inhibition is long, an overall poisoning

←

Figure 8.6. The diversity of patterns on related molluscs. (a) Transitions from branching lines to triangles on *Lioconcha castrensis* . (b) Branching lines can form a dense meshwork (*L. ornata*). (c) Branching lines can terminate blindly or fade away (*L. hieroglyphica*). (d) A specimen of *L. hieroglyphica* that has formed branched lines at younger stages (top) but larger triangles consisting of fine parallel lines at later stages. (e) Triangles and crossings superimposed. Pigmentation can be maintained for relatively long time intervals. (f) In *Lioconcha lorenziana* pigmentation remains at an elevated level after a pulse. The oblique lines have a fuzzy lower border.

occurs. (In contrast, if the time constant of this rapidly diffusing inhibitor is short, crossings can occur, see Figure 5.8).

After some delay a spontaneous trigger of new activations occurs preferentially in the gap left untouched by the two terminated waves (arrows in Figures 8.6a, c and 8.10). This seems to be curious behaviour. On the one hand, the inhibition is so strong that wave termination takes place even without collision. On the other hand, a spontaneous activation is possible somewhat later in the same cells that could not be triggered earlier by activated neighbouring cells. A very long-lasting, non-diffusible inhibitor must exist that provides a measure of the time since the last activation. Only when most of this inhibitor has decayed is spontaneous activation possible again. (The diffusible inhibitor is almost completely gone by this stage). If this inhibitor were diffusible, information on the location of the gap would be lost. The long-lasting inhibitor also has another function. Similar to the discussion in chapter 5, local accumulation of this inhibitor also forces pigment activation to move into a neighbouring region even though this region is already poisoned by the diffusible inhibitor. By itself, the rapidly diffusing inhibitor would lead to spatially stable patterns, in contrast to observations. The pushing effect of the non-diffusible inhibitor and the poisoning effect of the diffusible inhibitor together may cause a situation in which activation can no longer be maintained.

Considering all these features of a pattern of fine branching lines that terminate but re-trigger spontaneously in the gaps, at least three inhibitions must be involved (Figure 8.8):

(i)   A very rapid and non-diffusible inhibition that causes activation to last for a short time interval only. It determines the thickness of the oblique lines. Its diffusion (if any) must be much smaller than that of the activator; otherwise no travelling waves would form.

(ii)   A rapidly diffusing inhibitor with a long time constant whose overall accumulation poisons a region. This causes wave termination when two waves approach.

(iii) A very long lasting, non-diffusible inhibition responsible for the shift of activation despite the diffusible inhibitor. It also makes spontaneous activations possible again after an interval of non-activation. This occurs preferentially in the gaps left by earlier waves.

Several of the shells display a second pattern at lower pigmentation densities. In Figures 8.6c and 8.8 this faint pattern maintains the character of the main pattern while in Figure 8.6f it has more of an area-filling character. It is not clear whether this background pattern results from a partially independent system that forms in

Figure 8.8. Interrupted lines and branches. (a) *Lioconcha hieroglyphica*. (b) The highly diffusible inhibitor (red) causes termination of one or both approaching waves. After the decay of the long-lasting, non-diffusible inhibitor (blue-green), a spontaneous new activation is possible. It occurs preferentially in the gap left by two terminated waves; [S88].

Time →

Position →

a deeper layer of the shell, or whether pattern formation can proceed at such a reduced level. The systematic decrease of pigmentation levels in Figure 8.6c, for instance, suggests the second possibility. If this is true, the next question would be how pigmentation can spread with the same speed even though activation is reduced.

A system with three inhibitory effects has interesting features of its own and the resulting pattern elements can be recognized on the collection of shells given in Figure 8.6. That crossings become possible due to a diffusible inhibitor was discussed earlier. The third and non-diffusible inhibitor allows spontaneous activation to occur whenever a large non-pigmented region appears (Figure 8.9a). This is a feature clearly necessary for *L. hieroglyphica* (Figure 8.6c). If, in contrast, the second inhibitor is much less diffusible and has a longer time constant, neighbouring regions are much less affected and pigmentation slowly fades away. The oblique lines have a sharp upper boundary but a fuzzy lower one (Figure 8.9b), a feature characteristic for *L. lorenziana* (Figure 8.6f).

Other specimens of *Lioconcha hieroglyphica* display triangles that have a fine structure of parallel lines (Figure 8.10). Of course, the mechanisms leading to one or the other pattern within the same species cannot be very different. This is underlined by an even higher degree of similarity at younger stages. One possibility for the generation of these fine lines, therefore, is a less effective extinguishing reaction. If sufficient activator remains when the extinguishing reaction is over, an immediate re-activation is possible. The periodic extinguishing reaction shapes the fine structure. Triangles result since each newly reactivated area is a bit larger due to survival and spread at the margins.

In the specimen shown in Figure 8.10, the lower edges of several triangles are formed at the same time. Obviously, termination of triangle formation can be a global event (see chapter 6). The simulation assumes the presence of a hormone produced by all pigment producing cells. An increase in this globally distributed hormone makes the system more sensitive to the actions of the extinguishing reaction. As discussed above, if, by chance, all activations terminate, a spontaneous new trigger occurs preferentially in the gap between two triangles since the concentration of the non-diffusible inhibitor with the long time constant is lowest in this region.

Although the main features are reproduced, there are still discrepancies. In natural patterns, many more parallel lines can appear in series, forming larger coherent triangles. This causes problems in the simulations since, due to the accumulation of the long lasting inhibitor, the centre is more sensitive to the extinguishing reaction. But the very pronounced feature of spontaneous activations in the gaps left by preceding triangles (Figure 8.10) indicates that such an inhibition must exist.

The occurrence of pigment termination at the outer edge without previous termination at the centre suggests that the reaction causing the coincident termination of many triangles is much stronger than the usual oscillating reaction that

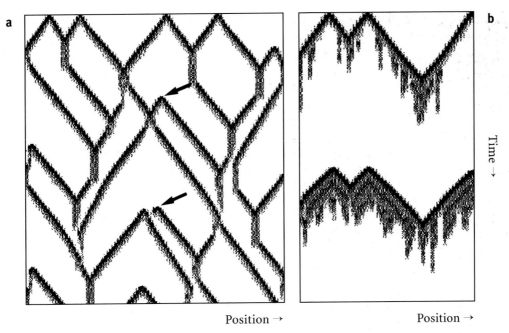

Position →   Position →

Figure 8.9. Patterns generated by a system with three inhibitions. (a) Formation of crossings *and* spontaneous initiations of two diverging lines. (b) Oblique lines with a sharp upper but a fuzzy lower border. These pattern elements can be recognized in species of the *Lioconcha* family (Figure 8.6) [S89a, S89b].

causes the fine parallel lines. An interaction between two oscillating systems is conceivable. The first system would produce the fine lines. Interference with the second system is decisive where termination occurs (see Figure 4.14). Occasionally this can be so strong that regions at the centre and at the tip of a triangle are affected simultaneously. In the next chapter, an alternative model for the generation of fine parallel lines with interruptions will be explored. At present it is difficult to determine whether these models will be helpful in overcoming the problems.

8.4 Breakdown as a failure of the enhancing reaction

Shells of *Voluta vespertilio* (Figure 8.11) typically have large but irregularly pigmented areas that are partially connected by oblique lines. The lower edges of these lines usually have a fuzzy appearance. In large pigmented regions a stripe-like modulation is frequently visible. These are very similar to the pattern elements discussed above for *L. hieroglyphica* (Figure 8.10). The pattern varies considerably among shells of the same species. To illustrate an alternative possibility, these patterns are simulated not by an explicit extinguishing reaction but by an enhancing reaction that brings the system close to a steady state. This simulation uses the same interaction as the meshwork pattern shown in Figure 7.8, although the patterns look very different. If this mutual stabilization causes a situation in which both the pigmentation and enhancing systems are on the border of maintaining

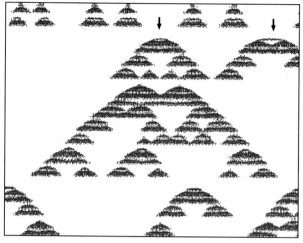

Time →

Time →

Position →

Figure 8.10. *Lioconcha hieroglyphica*: large triangles with a fine structure. The simulation is based on a two inhibitor model (Equation 5.4) plus a hormone for simultaneous termination of many triangles. Oscillations cause the fine structure of triangles [S810].

a steady state, minor fluctuations can become amplified over the course of time until a breakdown occurs. Since the enhancing reaction is assumed to spread more rapidly, this breakdown can occur over a large region due to the synchronization of these fluctuations. Pigmentation eventually survives in marginal zones by forming travelling waves. After variable time intervals the stabilization again becomes sufficient to generate steady state activator production. The simulations include only the pigment and enhancement reactions (Equation 7.1 a-d, page 113). In this simple form it is relatively parameter-sensitive. This corresponds to the large variations in observed patterns. The sensitivity disappears if additional regulatory processes such as those discussed above are included, for instance, global control or long lasting inhibitions. This model introduces a different possibility for large scale breakdown. The extinguishing reaction is replaced by the failure of a rapidly spreading enhancing reaction.

8.5   Conclusion

Triangles underline once more the intimate coupling between the extinguishing and enhancing effect of a modifying reaction. In the case of triangles, the extinguishing reaction is followed very rapidly by the enhancing reaction, causing rapid recovery after a large scale breakdown. In many molluscs shell growth is not a continuous process but occurs in phases, leaving distinct growth lines. The coincidence between these lines and the lower edges of the triangles (as seen in Figure 8.5b) suggests that the oscillations required for triangle formation may be part of a general physiological process taking place in the molluscs.

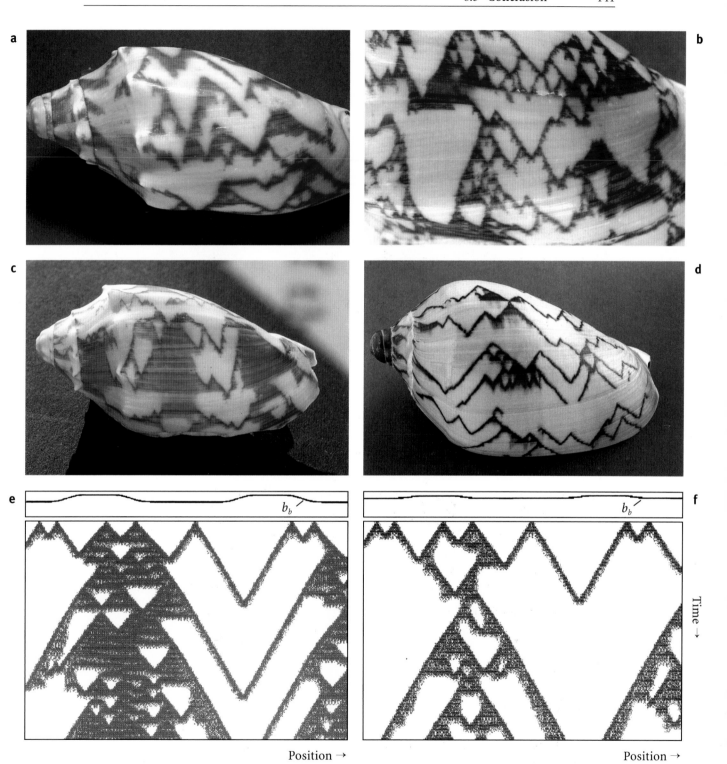

Figure 8.11. Irregular arrangement of large pigmented areas and isolated oblique lines. (a-c) *Voluta vespertilio*, (d) *Voluta nobilis*. (e, f) Model: the enhancing pattern is not restricted in space but is on the border of oscillations. Over large time intervals mutual stabilization is sufficient to maintain a steady state. Minor fluctuations can amplify over the course of time such that large scale breakdowns occur. If mutual stabilization is interrupted, it may take some time before it is established again. The simulations are performed by using equations 7.1 a-d. To simulate the two darker bands visible on most shells, substrate production in the pigment reaction is assumed to have some modulation ($b_b(x)$). [S811e, S811f].

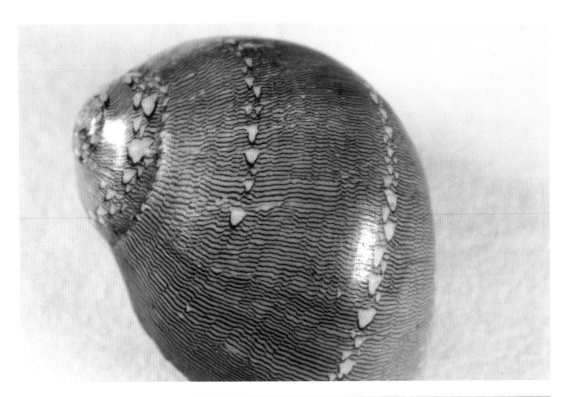

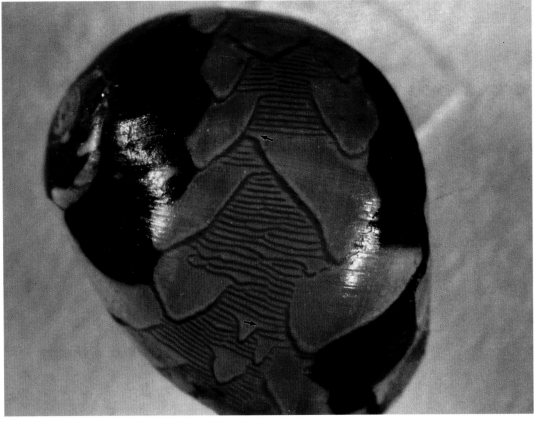

Chapter 9

Parallel lines with tongues

The upper shell in Figure 9.1 is decorated with many fine parallel lines. This pattern suggests the same synchronous oscillations as described earlier (see Figure 3.4). However, at particular positions, the parallel lines are deformed into U- or V-shaped gaps. The pattern on the lower shell is based on the same principle; only the size and regularity of the gaps are different. On the upper shell the gaps are restricted to particular positions. On the lower shell two broad bands are nearly free of parallel lines while smaller gaps appear at more scattered positions. The shells belong to the species *Clithon oualaniensis* (in older literature also termed *Neritina* or *Theodoxus oualaniensis*). These small brackwater snails are frequent on shores around India and Sri Lanka and display an incredible richness of patterns. Grüneberg (1976) made a careful study of the polymorphism of these shells. He termed the deviation from parallel straight lines "tongues". His article contains many examples of different types of patterns, transitions from one type to another, and pattern regulation after perturbation.

What is the basis of tongue formation? Obviously, a tongue results from the temporary suspension of otherwise almost synchronous oscillations. A non-pigmented area results that is re-pigmented from neighbouring regions in which oscillations survived. The V- or U-shaped oblique lines that border the tongue indicate that travelling waves move into the region of interrupted oscillation, filling the gap. After the waves meet, normal oscillation usually continues, at least for a certain interval.

Figure 9.1. Parallel lines with tongues: Patterns on *Clithon* (or *Theodoxus*) *oualaniensis*. The parallel lines indicate synchronous oscillations. These may be interrupted both in smaller (top) or larger regions (bottom). The resulting gaps, the "tongues", are slowly filled by waves that have spread from regions in which the oscillation survived.

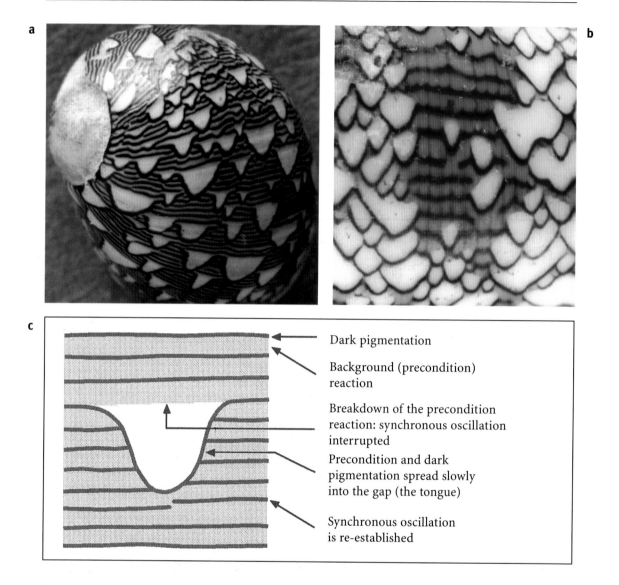

Figure 9.2. Tongues, a temporary interruption in synchronous oscillations causing gaps in a pattern of parallel lines. (a) Details of *Clithon oualaniensis*. A grey background pattern is visible between the parallel lines. It is absent in the tongues but remains active for a while after the last pulse at the beginning of a tongue. As far as can be seen, the background system switches off abruptly, suggesting that its survival does not depend on the half life of a substance but results from the property of a non-linear reaction. (b) The same pattern elements in the more complex pattern of *Conus textile*. The beginning of a tongue is more variable in respect to the last unaffected line. In addition, finer oblique lines with branches are formed. (c) Schematic drawing of a possible mechanism that leads to the formation of tongues. Two systems are superimposed, an oscillating system that forms dark parallel lines and the background system. Oscillations are only possible when the background pattern is present. The oscillating system may have a role in maintaining the background system. A tongue would be formed when the oscillating system refreshes too late.

9.1 Survival using a precondition pattern

An important hint in determining the underlying mechanism comes from a background pigmentation that is visible between the parallel lines on some shells (Figure 9.2a). This is most clearly visible on parts of the shell of *Conus textile* (Figure 9.2b) because of the light brown pigmentation that is absent in the tongues. Thus, three different levels of pigment production can be distinguished: a high level causing the dark pigmented lines, a lower level causing background pigmentation between the lines, and a zero level inside the tongues. Shells with three distinct levels of pigmentation have been discussed earlier (Figure 7.6 and 7.9) and a case has been made that this involves two pattern forming systems. The present pattern suggests that the background system acts as a precondition. It must be in the "ON" state in order for the pigmentation system to oscillate, forming the dark parallel lines. Whenever the precondition disappears locally, for whatever reason, the oscillations are interrupted (Figure 9.2c). It is worth mentioning that, in contrast to the shell shown in Figure 9.2a, the background pattern is not visible on the bottom shell of Figure 9.1. Reconstruction of the underlying mechanism of such shells would be much more problematic without hints from related specimens.

According to this view, the present pattern is similar to the complex patterns with branching discussed in chapter 7 in which the pigment system is modified by a second auxiliary system. The difference, however, is in the function of the auxiliary system. In the former case, branch and stalk formation result from a temporary shift of the pigmentation system from an oscillating mode to a temporary steady state under the influence of an enhancing system: survival by locally elongated pigment deposition time. In contrast, the parallel lines discussed here result from a rapid re-triggering of the pigmentation system under the influence of a precondition system: survival by revival. Minor alterations in the mechanism are sufficient for this change. It may be that during the diversification of the species the need of an enhancing pattern became so strong that it began to act as precondition, i.e., that without the enhancing pattern no pigmentation was possible. While in branch formation the pigmentation system is (presumably) a precondition for the enhancing system, in the present pattern the background pattern between the lines indicates that the precondition can survive for a certain period without the pigmentation system. The coexistence of both parallel lines with tongues and oblique lines with branches on the shell of *Conus textile* (Figure 9.2b) suggests that a nearly continuous transition from one mode to another exists.

The simultaneous formation of two different elementary pattern elements, lines parallel and oblique to the growing edge, is another indication of the involvement of two patterning systems since they require conflicting sets of parameter values. Parallel lines result when synchronization is enforced by a strong coupling between neighbouring cells (see Figure 3.4). In contrast, oblique lines are caused by a moderate diffusion of the self-enhancing agent combined with an almost non-diffusible antagonist. Slow infection of neighbouring cells and thus travelling waves are the

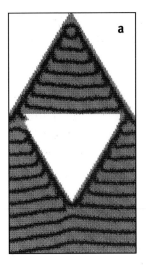 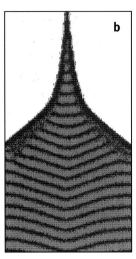 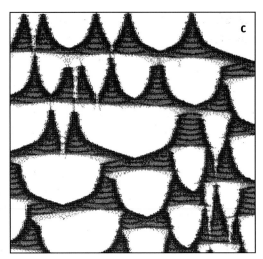

Time →

Position →

Figure 9.3. Models for parallel and oblique lines on a single shell. A background system (green) forms the precondition for the oscillating system (black). (a) The substrate of the oscillating system is only produced when the background system is active. The oblique lines of the pigment system follow the slow spread of the background system. Their inclinations are thus determined by this spread. Repairing a gap in the background pattern (artificially introduced in this simulation) leads to tongue formation. (b) The precondition pattern is required as a co-factor for dark pigmentation. (c) A model that can be excluded: Gaps in the background system are produced by a separate extinguishing system (red). This leads to interruptions in pigment lines, in contrast to the actual observation; [S93a, S93c].

consequence (see Figure 3.7). This conflict disappears if two systems are involved. The precondition/background system spreads slowly causing the time-consuming filling of the tongue. In contrast, the rapid spread of the dark pigmentation system leads to a synchronization of the oscillations. Despite the fact that its spread is rapid, it cannot invade the tongue immediately because the precondition must be re-established first.

If two patterns are superimposed, it is difficult to determine whether both reactions are directly involved in pigment production or, as was assumed in chapter 7, only one reaction produces pigment and the second causes its enhancement. In the case of parallel lines with tongues background pigmentation may result from a baseline activation required for the rapid triggering of the oscillating system. It is naturally restricted to regions in which the pattern is active.

9.2 Tongue formation: refresh comes too late

This section will describe several models but will conclude that the shape and arrangement of the tongues are very restrictive for possible mechanisms. In Figure 9.3a it is assumed that the substrate of the oscillating reaction is only produced when the background reaction is in an active state. The background pattern is, therefore, a precondition for the sustained oscillations and can spread slowly into non-activated regions. A breakdown in the background pattern leads to an interruption in the oscillations and the gap is filled in by oblique pigmentation lines. With such a model one would expect a certain time lag between the re-activation of the background pattern and the onset of dark pigmentation, since the substrate for the oscillatory reaction must be produced first. Grüneberg (1976) concluded from his microscopic inspection of *Clithon* shells that two pigmentation systems are involved: a leuco-system that produces a white pigment and a melano-system that produces the dark pigment. He observed that the activity of the leuco-system always precedes that of the melano-system. In Figure 9.1b a lighter pigmentation is clearly visible in front of the dark oblique lines and may be attributed to the second pigment system. In contrast, in *Conus textile* (Figure 9.2b) the spread into the gap is headed by a dark line, not by background pigmentation. This suggests that oscillations can be triggered immediately once the precondition is re-established. This is the case when the precondition pattern acts as a co-factor. Figure 9.3b provides a corresponding simulation in which, in addition, the background system is assumed to be non-diffusible. Its spread results from a cross-reaction with the oscillating system. Due to its rapid diffusion, the oscillating system spills over the ON-OFF border of the background system. The cross-reaction triggers activation of the precondition system and allows it to extend into a formerly non-activated region. This system provides a better description of the acceleration of the waves, i.e., of the rounded shapes of the tongues.

The cause of the breakdown of the background reaction is a crucial problem. One suggestion would be to assume a separate extinguishing mechanism that would lead to gaps in the background pattern, similar to the simulation in Figure 7.2. However, the corresponding simulation (Figure 9.3c) shows clearly that this cannot be the case. The sudden de-activation of the precondition/background pattern would lead to interruptions in the parallel lines, a feature that is rarely observed on real shells. Or, conversely, the fact that the beginning of a gap is usually parallel to a dark line, suggests that the oscillatory system is involved either in maintaining or switching off the precondition system.

One could imagine that the oscillating system has an extinguishing influence on the background pattern, and that sometimes the background pattern does not survive. However, especially in patterns of the *Clithon* type, the background pattern remains in the ON-state for some time after the last oscillation (Figure 9.2a). If the oscillating system had an extinguishing influence, the probability of killing the background pattern would be highest when the oscillating system reaches its highest level. This is not the case. An overshoot of the antagonistic reaction is equally unlikely since this would also appear immediately after a pulse.

The lengthy survival of the background pattern suggests another mechanism: the oscillating system may act to refresh the background system. When a pulse of the oscillating system appears too late, the background system dies out along with the prerequisite for sustained oscillations. After each pulse, a type of race takes place to determine whether the next pulse will arrive early enough to accomplish the refresh. The mutual dependence of a steady state system and an oscillating system may be illustrated with an analogy. The living state of a higher organism depends on a beating heart. But the heart can only beat if the organism is alive. If a single heart beat comes too late, the organism may die. Of course, if the organism dies for other reasons, the heart will also stop beating.

The repetitive breakdown of the background system suggest that the latter is close to the border of stability. On some shells the tongues are arranged in a scattered way, on others they appear at more staggered positions. In the first case, their initiation may depend on statistical fluctuations. In the second case, however, a signal must be present that establishes a time interval between the close of one tongue and the trigger for the next tongue. During tongue formation a substance may be produced that stabilizes subsequent oscillations. Or, either the oscillating or background system may produce a poison that eventually terminates the precondition system early, making the next breakdown more likely. A decrease in oscillation frequency would have the same effect. Occasionally somewhat larger spacing between the parallel lines is visible at the onset of tongue formation (arrow in Figure 9.1) which is an argument in favour of the latter possibility.

Modelling the generation and action of this tongue-inducing signal turns out to be very difficult. With most of the models envisioned poisoning would lead to longer and longer intervals between pulses (larger and larger distances between lines) until breakdown occurs. In terms of the heartbeat analogy, dying does not happen from one heartbeat to the next. In shells, however, breakdown is not announced by a preceding modification in the oscillating pattern, except for the possible minute increase in the line spacing mentioned above. The build up of the signal must proceed in a hidden way – a puzzling feature that was also required in earlier simulations (chapter 7). There does not seem to be any visible structure to the background pattern. If the probability for breakdown increases over time one would expect the background pigmentation to have a graded profile. This is, however, not the case.

The delay between the last pulse and the breakdown of the background system suggests that the refresh function lasts longer than the pulse width of the pigmentation reaction. Oscillations require a longer time constant for the antagonist (substrates or inhibitors). Thus, it is presumably the antagonist and not the activator of the oscillating pigmentation system that is involved in refreshing the background system.

The simulations in Figures 9.4 - 9.6 have been made with equation 9.1 (page 157) which assume that the background/precondition system is generated by an activator-inhibitor interaction (green in the simulations). An additional long-lasting inhibitor functions as the poison mentioned above. It is produced whenever the precondition system is active. Its accumulation initiates tongue formation while its decline during the tongue-phase has a stabilizing effect on the background system and thus on the subsequent synchronous oscillations. The precondition

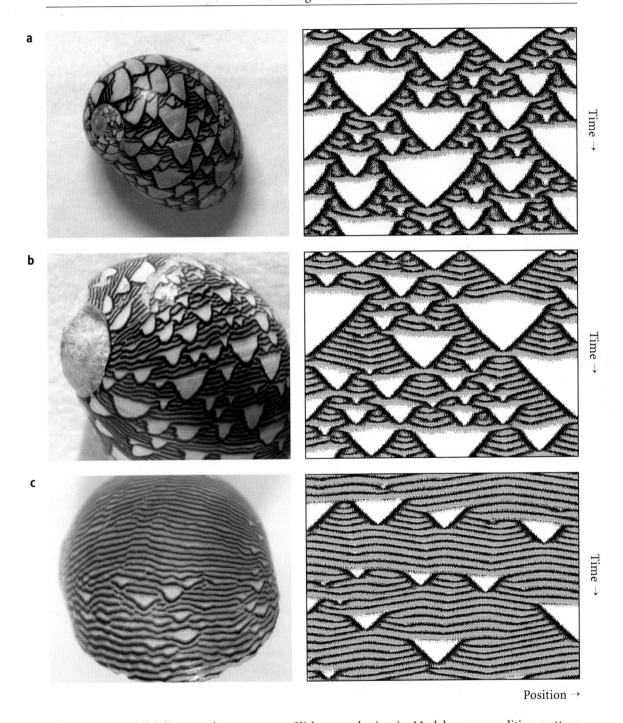

Figure 9.4. Parallel lines and tongues on *Clithon oualaniensis*. Model: a precondition pattern (green) is required in order for synchronous oscillations to take place (dark parallel lines). The oscillation system acts to refresh the precondition pattern. Due to the accumulation of an inhibitory metabolic product, a pulse may arise too late to maintain the precondition pattern. Non-pigmented regions emerge that are filled by travelling waves. The frequency of these tongues depends on sensitivity against the poisoning product. Calculated using equation 9.1; [S94].

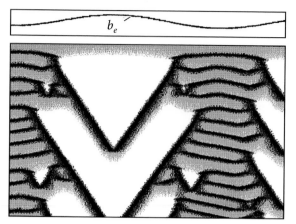

Time →

Position →

Figure 9.5. The restriction of the parallel lines into bands, a characteristic feature of many *Clithon* shells. These bands are frequently connected by oblique lines. Model: in regions with a higher production rate of the poison (b_e in equation 9.1), activation of the precondition pattern is too short to allow several oscillations. Isolated oblique lines (travelling waves) result that have their origin in one of the parallel lines. Note that these lines do not branch [S95].

system is close to the border of stability. Therefore, small changes that do not influence the overall pattern are sufficient to switch it off. For the oscillating pattern (black), an activator-substrate interaction is employed. In the simulations an activator-substrate system is more convenient for both the spread into the tongues and the synchronization. The precondition system is required as a co-factor for the oscillating system. The refresh function of the oscillating system has been implemented in the following way: In an activator-substrate model, the substrate concentration declines dramatically with each pulse and recovers slowly afterwards (see Figure 3.3, page 43). The substrate is assumed to have a cross-inhibitory influence on the precondition system. This inhibitory effect is, therefore, very low immediately following a pulse and increases thereafter. With time, the stability of the precondition system becomes more and more endangered. Either the next pulse arises in time to refresh the system or a sudden breakdown occurs. From this model it is expected that the breakdown would occur very shortly before the next pulse is due to appear, which agrees with the natural pattern.

In many specimens of *Clithon oualaniensis* the parallel lines are restricted to three or five bands. These bands may be connected by oblique lines (Figures 9.5 and 9.1b). According to Grüneberg (1976), the separation of a coherent field of parallel lines into bands occurs more frequently in older animals. The bands are created by small tongues formed in rapid succession at particular positions (see Figure 9.1a) and becoming larger and larger. This process is not necessarily irreversible. In the model, isolated oblique lines suggest that the period in which the precondition is active is too short to host more than one pulse of the pigmentation system (Figure 9.5). In other shells, the inhibition is so strong that even these

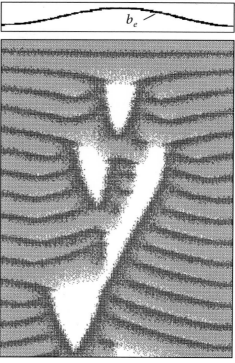

Time →

Position →

Figure 9.6. Travelling gaps in *Neritina virginea*. When the dark system is not directly triggered after gap formation, the breakdown of the precondition system may spread (arrows). Oblique bands free of fine parallel lines may result [S96].

connections are no longer formed. Examples were given earlier in another context (see Figure 1.10).

In *Clithon*, isolated oblique lines never branch, in contrast to the situation discussed in chapter 7. According to the model, the oblique lines in *Clithon* result from a wave-like spread of both the precondition *and* the pigmentation system. In contrast, in patterns with branching, the oblique lines are formed by one system only, while the occasional triggering of the second system may cause the formation of a branch (compare the black-green line in Figure 9.5 with the green dots along black lines in Figure 7.8). In the present pattern, when the precondition remains active for a long period, the result is not a branch but two successive lines or small patches of parallel lines (see Figure 9.7f).

After tongue initiation, the next dark pigmentation is usual involved in filling the gap immediately. Occasionally, however, several oscillations are required before a travelling wave (oblique line) is formed (arrows in Figure 9.5). The formation of a wave seems to be correlated with the spread of the oscillations, but it is difficult to determine which is the cause and which is the effect. On some shells a gap can be formed that enlarges on one side and shrinks on the other moving across the field (Figure 9.6). This phenomenon is reproduced in the simulation. The moving gaps shed light on the interplay between the two systems. On its own, the precondition system may have a tendency to retract. In contrast, the pigmentation system has a tendency to move into the gap since more substrate is available there. Gap formation and its closure is thus a fight between the retraction of the precondition system and the spread of the oscillating pigmentation system, pulling the precondition system with it by its refresh function. Accordingly, the probability of enlarging or travelling gaps increases with a decrease in the rate of substrate diffusion, since the pigmentation system's attraction to the gap is lower. As in the real pattern only the lower side of such an oblique gap is bordered by a dark oblique pigmentation line. The movement of the gap terminates when the dark system is also triggered at the crumbling side, causing a bilateral spread into the gap and thus its closure.

9.3   Variations on a common theme

Related species display a large variety of patterns. Examples are given in Figure 9.7, although detailed modelling of this diversity has not been accomplished. The shell in Figure 9.7a displays round non-pigmented patches on a more densely pigmented background. This pattern may appear as a complement to the staggered dot pattern in Figure 5.11. However, the patches are less regularly arranged and have a dark crescent-shaped line at their lower edge. On closer inspection the impression of a background free of textures emerges because the parallel lines are extremely fine and very close to each other. The shell in Figure 9.7b displays a similar pattern with a somewhat larger spacing between the lines. The white drops are tongues and the dark crescents result from an overshoot of the pigment reaction.

One remarkable feature of the shell in Figure 9.7a is the long faint shadows under the white patches (arrow) which are distinctly different from the overshoot phenomenon. This suggests that when pigment deposition is interrupted, background pigmentation becomes enhanced for an interval at least as long as the non-pigmented period. This enhancement suppresses the onset of tongue formation and is thus responsible for the staggered position of the white patches. New patches can appear only outside these shadows. According to the model, the shadows result from the stabilizing influence of the gap on successive oscillations. In the example given in equation 9.1 the stabilization is caused by a decline in poison concentration during tongue formation. Since a shadow does not change its lateral extension over the course of time, the poison must be non-diffusible. To obtain the rounded shape in the upper parts of the patches the breakdown of the background pattern must spread, in contrast to many patterns discussed earlier (but see Figure 9.6). Moreover, the nearly circular shape of the patches requires the breakdown to spread with about the same speed as the re-population of the

Figure 9.7. Variation in patterns of the *Clithon* type

gaps. Since both processes presumably depend on different mechanisms, these equivalent speeds are non-trivial.

In Figure 9.7c the oblique lines that border the tongues are much broader, suggesting either a strong overshoot or an active process (see Figure 7.4a). In Figure 9.7d, only the oblique lines that border the tongue are darkly stained, while the closely spaced parallel lines are pale. The oblique and parallel lines in Figure 9.7e have also different colours, black and red. In *Puperita pupa* (Figure 9.7f) two oblique lines frequently appear close together, followed by a larger region free of pigment. In terms of the model, this suggests that the precondition reaction is active for a relatively short period in which only two pulses of the oscillating pattern can fit. Transition into patches with parallel lines are also clearly visible. In the model this would occur when the precondition pattern is active for a longer period.

9.4   *Conus textile:* tongues and branches on the same shell

The highly poisonous snail *Conus textile* forms much larger shells (up to 7cm) that are decorated with similar pattern elements. Some examples were given earlier (Figure 1.12 and 9.2). Figure 9.8 shows another specimen. The dark parallel lines, the light brown background pigmentation and the tongues are usually restricted to two bands. The interesting regions are between these bands. Fine lines are formed that occasionally branch. Therefore, these patterns provide a link between the complex branching patterns discussed in chapter 7 and the parallel lines with tongues discussed above. The occurrence of branches suggests the use of the model developed earlier (equation 7.1). In the simulation the dark bands with parallel lines and the oblique lines with branches are reproduced using this model (Figure 9.8).

What are the common features in this and the *Clithon* patterns, and what causes the differences? According to the models, one difference is the pattern that oscillates. In the *Clithon* model, the pigmentation pattern oscillates while the auxiliary (precondition) pattern remains in a steady state as long as no tongue is formed. The latter pattern provides the ingredients for a rapid re-trigger. In the branching model used for the simulation of *Conus textile* (Figure 9.8) the reverse is true. The pigment pattern may obtain a steady state while the auxiliary (enhancing) pattern oscillates. This leads to large pigmented areas with parallel stripes of enhanced pigmentation. If the pigment pattern does not remain in a steady state, travelling

$\longrightarrow$

Figure 9.8. Tongues and branches on the same shell: *Conus textile*. The simulation uses a pigment system and an enhancement system (equation 7.1a-d). The first system produces the background pattern or the travelling waves depending on the rate of substrate production b_b. The enhancing pattern is responsible for the dark lines on top of the background system or the branching of oblique lines, depending on whether the pigmentation system is in a steady state or is producing travelling waves. For arrows see text; [S98].

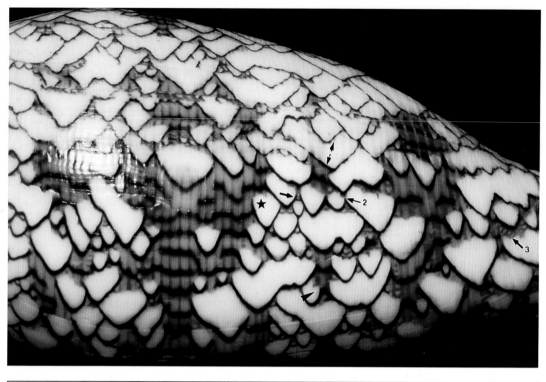

waves are formed and the periodic enhancement causes branching. In the *Clithon* model, oblique lines result from joint waves of both the pigment and auxiliary system. No branches can be formed since, during this spread, the auxiliary pattern is already in the ON-state. The background pigmentation results from residual activation of the pigment system by the auxiliary system. In contrast, the background in the branching model results directly from the pigment reaction. This provides a rationale for the much stronger background pigmentation on shells like *Conus textile* or *Conus episcopus* (see Figure 7.6) in contrast to shells of the *Clithon* type.

As shown in Figure 9.8 the branching model correctly describes the parallel lines, which are much thicker than the oblique lines. This results from the influence of the enhancing pattern which elongates the period of pigment deposition. This was a necessary condition for branch formation. In this view, branch formation and the thick lines are two manifestations of the same interaction.

On the other hand, *Conus textile* displays many features that cannot be explained by the branching mechanism but suggest the actual mechanism is closer to that developed for *Clithon*. For instance, the oblique lines have variable thickness and degrees of pigmentation. This is most obvious when an oblique line has a light-brown band at its lower edge. The dark parts of these lines are distinctly thicker than lines without these bands (compare the two lines marked by a double arrow in Figure 9.8). This suggests that joint waves from the pigmentation and the auxiliary system are possible and that the trigger of the latter does not necessarily lead to branch formation. It is very remarkable that increased intensity does not lead to waves with a higher velocity.

Furthermore, branch formation is not the only mode of generating new waves within the white region (see symbols in Figure 9.8). Occasionally the light-brown background pattern is triggered and remains active long enough to trigger the dark system again (arrowhead and arrow). Therefore, *Conus textile* shows both modes of survival on the same shell: elongation of lifetime (branching) and revival (a rapid re-trigger *via* a second system). Often the trigger of the oscillating system occurs somewhat earlier at the border between a white and light-brown region (asterisk). Thinner light-brown bands cause the next trigger to occur sooner (arrow 2). Obviously, proximity to a non-pigmented region can accelerate the next pulse of the oscillating system.

The background reaction cannot spread on its own. On the contrary, it may rapidly retract (arrowhead), a situation analogous to that of *Clithon* (Figure 9.6). The white triangles or tongues, are described much better by the *Clithon* model and retraction is certainly an integral part of tongue formation in *Conus textile*.

The tongues in *Conus textile* are much less regularly arranged than in *Clithon*. This makes it difficult to design models for their initiation. Moreover, this initiation does not have a strict phase relation with respect to the last dark line. The breakdown may also appear immediately after the dark line. Therefore, in contrast to *Clithon*, the dark lines in *Conus textile* presumably serve no maintenance function.

Equation 9.1: Parallel lines and tongues

The pigmentation system (an activator - substrate system) in which the pre-condition system c acts as co-factor:

$$\frac{\partial a}{\partial t} = s\,c\,b\,a^{*2} - r_a a + D_a \frac{\partial^2 a}{\partial x^2} \qquad (9.1.a)$$

$$\frac{\partial b}{\partial t} = b_b(x) - s\,c\,b\,a^{*2} - r_b b + D_b \frac{\partial^2 b}{\partial x^2} \qquad (9.1.b)$$

The precondition system (an activator - inhibitor system):

$$\frac{\partial c}{\partial t} = r_c \left(\frac{c^{*2}}{d + c_c b + s_e e} \right) - r_c c + D_c \frac{\partial^2 c}{\partial x^2} \qquad (9.1.c)$$

$$\frac{\partial d}{\partial t} = r_c c^{*2} - r_d d + D_d \frac{\partial^2 d}{\partial x^2} \qquad (9.1.d)$$

$$\text{with} \quad a^{*2} = \frac{a^2}{1 + s_a a^2} + b_a \quad \text{and} \quad c^{*2} = \frac{c^2}{1 + s_c c^2} + b_c$$

The production rate of the metabolic product, the "poison":

$$\frac{\partial e}{\partial t} = b_e(x)\,c - r_e e \qquad (9.1.e)$$

$c_c b$ The substrate of the pigment system has an inhibitory influence on the precondition pattern. It is at its lowest immediately after a pulse and increases thereafter.

$s_e e$ The inhibitory influence of the poison on the precondition reaction. Both inhibitory effects bring the precondition system c to the border of stability.

$b_e(x)c$ The production rate of the poison depends on the presence of the background system. Either this production rate or the sensitivity s_e may be space dependent, producing large tongues at regular distances (Figure 9.5).

The light-brown background is also modulated by stripes parallel to the direction of growth which may resolve into a fine meshwork-like pattern (arrow 3). In other specimen (see Figures 1.12 and 7.1) longer non-pigmented phases may give rise to branches in quick succession, creating a fine bubble-like pattern. The similarity between these patterns and the complex patterns discussed earlier (Figure 7.11) is easily recognizable.

9.5   Missing elements, missing links

The models developed in this book are certainly only first approximations. Many details of the patterns have not been described. Chains of white drops (Figure 7.9), irregular alternations between dark and white patches (Figure 1.9) and the transition into fine meshworks (Figure 7.11) are examples. However, these problems are discovered only when a theory is available. The models help to uncover elements that are not fully understood, and discrepancies become apparent only when models are formulated mathematically.

The last figure contains a collection of shell patterns that may stimulate further model building. *Conus marmoreus nigrescens* (Figure 9.9a) displays the same white drops on a black background as *Conus marmoreus* (Figure 7.2). However, the black region has a substructure of closely packed parallel lines and the white drops are unmistakable analogous to tongues. It is thus a missing link and suggests that the formation of white drops in *Conus marmoreus* (Figure 7.2a) and of tongues in *Clithon* may have the same basis. Unification of the corresponding models is certainly the biggest challenge in the future modelling of shell patterns.

Conus nicobaricus (Figure 9.9b) shows fine branching lines superimposed with cloudy patches of dark black-blue pigmentation. As far as one can observe through this cloud, the number of branches but not the thickness of lines increases tremendously in this regions. The non-linear models employed in the simulations tend to produce regions of all-or-none activations with sharp demarcations. How do the cloudy patches arise? How do they influence the probability of branching?

Conus vicweei (Figure 9.9c) displays the usual oblique lines but the lines are white on a pigmented background. Does this result from a normal reaction that either causes the deposition of white pigment or suppresses (extinguishes!) the pigmentation reaction locally? Alternatively, are reactions conceivable in which the self-enhancing process leads directly to a travelling local collapse of pigment production?

Many specimens of *Oliva porphyria* have a second pattern in the background. This is most clearly seen between the two arrows in the specimen in Figure 9.9d.

Figure 9.9. Some of the many patterns that await modelling

a

b

c

d

e

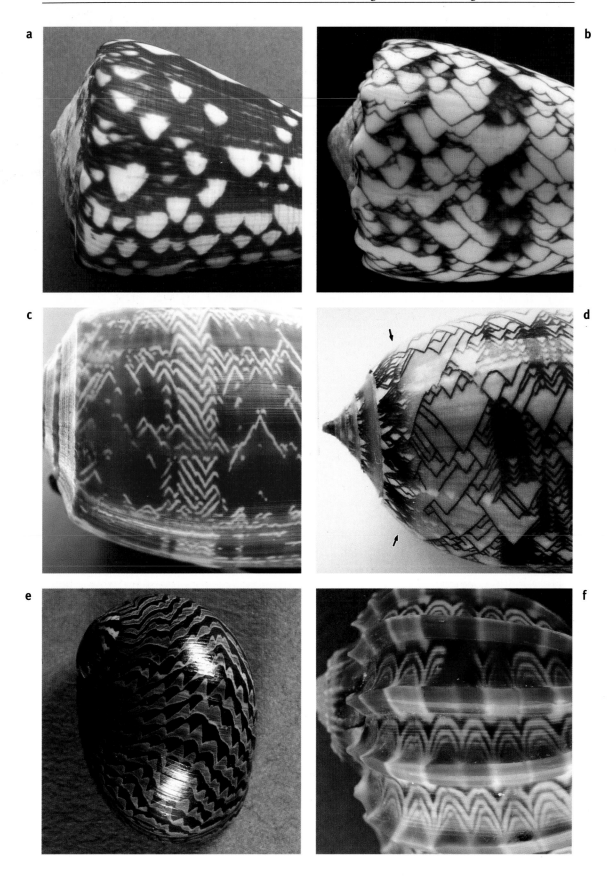

f

The two patterns merge into each other without much interference. Normal background pigmentation obviously results from this underlying system. Is this caused by another layer of cells in the mantle gland that behaves similarly but is partially uncoupled? Furthermore, what is the origin of the columns of densely packed parallel lines on this shell? Does this have a similar basis to the dense branching underneath the dark patches in Figure 9.9b?

Other patterns await models as well. In *Neritina communis* (Figure 9.9e) the oblique lines have a zigzag pattern. How does this arise? Is it a smoothed chain of triangles? The last example, *Harpia major* (Figure 9.9f), shows a regular pattern of parallel ribs. After each rib a space-dependent oscillatory pigment deposition takes place. The first stroke is the most vigorous; further strokes are more damped until a transition into steady state pigmentation occurs. Soon after this stage is reached, a new rib is formed. This oscillation is modulated by a spatial pattern, showing that more complex superpositions of stable and oscillating patterns can be realized. These examples certainly do not exhaust the list of shells yet to be simulated but it is hoped that they will stimulate further attempts.

Epilogue

Stable patterns, oscillations and travelling waves are the basic building blocks of shell patterns. These phenomena have a common base: the coupling of self-enhancing and antagonistic reactions.

Shell patterns have been used to demonstrate the general principles of mechanisms that generate stable patterns, oscillations and travelling waves. By comparing these mechanisms with everyday experiences such as pattern formations caused by erosion, waves of influenza or the flickering of a candle, the universality of these principles has been illustrated. Of course, shell patterns are only a very special case of pattern formation in space and time. The same mechanisms also play a decisive role in more important biological and non-biological processes. The development of higher organisms from a single cell, the fertilized egg, requires a cascade of at least temporarily stable patterns. Travelling waves are crucial for the organized contraction of the heart muscle. Neither thinking or sensing would be possible without nerve pulses that travel along a nerve fibre. The patterns on sea shells may not be convenient for investigating the molecular basis on which pattern formation is based. But their unique feature of maintaining a historical record and their incredible diversity recommend shell patterns as a natural picture book for studying dynamic systems. I hope that a reader coming across a shell will start to read the pattern just like a fascinating book.

In his novel "Doktor Faustus", in the same chapter quoted in the preface, Thomas Mann describes Jonathan Leverkühn's thoughts on the creative irregularity and beauty of frost flowers. He speaks about the *"creative dreaming of Nature"*. And further, *"... what occupied him was the essential unity of the animate and so-*

called inanimate nature, it was the thought that we sin against the latter when we draw too hard and fast a line between the two fields, since in reality it is pervious and there is no elementary capacity which is reserved entirely to the living creature and which the biologist could not study in an inanimate subject."[1]

In this book I have tried to listen to the creative dreaming of nature and I hope I have contributed a little to a reconciliation between the animate and inanimate worlds.

------

[1]Thomas Mann, Doktor Faustus, translation by H. T. Lowe-Porter, Penguin Books, p. 23

Figure 10.1. Trophon shell. Inspiration for this sculpture-like view came from shell photographs by Andreas Feininger (see Feininger and Emerson, 1972)

Chapter 10

Shell models in three dimensions

Przemyslaw Prusinkiewicz and Deborah R. Fowler[†]

Department of Computer Science
The University of Calgary

[†]now at Hi Tech Toons

Inspired by the models of pigmentation patterns developed by Dr. Meinhardt, we pursued a further goal — to create a comprehensive model of seashells that would incorporate these patterns into three-dimensional shell shapes. Our motivation was twofold. On the one hand, in the absence of a formal measure of what makes two forms and patterns look alike, it is often necessary to rely on visual inspection when comparing models with nature (Prusinkiewicz, 1994). Realistic presentation adds credibility to such comparisons by removing potentially misleading artifacts. On the other hand, we consider visual simulations a celebration of nature's beauty similar to painting, sculpture, or photography (Figure 10.1). Our results are described here according to the paper (Fowler *et al.,* 1992).

10.1 Mathematical descriptions of shell shape: a brief history

The essence of shell shape is captured by the logarithmic spiral, characterised mathematically by Descartes in 1638 (see Thompson, 1952, page 754) and first applied to describe shell coiling by Moseley (1838). By the beginning of the twentieth century, the logarithmic spiral was observed in many artificial and organic forms (Cook, 1914). Thompson (1952) presented careful measurements of a wide variety of taxonomic and functional shell types, and showed their conformity with the logarithmic model.

The application of computers to the visualisation and analysis of shells was originated by Raup. In his first paper devoted to this topic, he introduced two-dimensional plots of longitudinal cross-sections of shells as a blueprint for manually drawing shell forms (Raup, 1962). Subsequently, he extended his model to three dimensions (Raup and Michelson, 1965) and visualised shell models as stereo pairs to emphasise the three-dimensional construction of the shells (Raup, 1969). His models were plotted as a collection of dots or lines.

In the pursuit of realistic visualisations, Kawaguchi (1982) enhanced the appearance of shell models using filled polygons, which represented the surface of shells more convincingly than line drawings. Similar techniques were used subsequently by Oppenheimer (1986), and Prusinkiewicz and Streibel (1986b). A different approach was adopted by Pickover (1989, 1991), who approximated shell surfaces using interpenetrating spheres. They were placed at carefully chosen distances from each other and rendered using periodically altering colours to create the appearance of a ribbed surface with stripes.

Recent work on the modelling of shells has been characterised by an increasing attention to detail. Illert (1989) introduced Frenet frames (see Bronsvoort and Klok, 1985; do Carmo, 1976) to precisely orient the opening of a shell. His model also captured a form of surface sculpture. Cortie (1989) allowed independent tilting of the opening in three directions, presented models with apertures defying simple mathematical description, and extended the range of surface ornamentations captured by the model.

10.2 Elements of shell shape

Our model of shell geometry is similar to that introduced by Raup and developed further by Cortie. The underlying ideas, however, were already present in the work of Thompson (1952, chapter XI). We quote his observations in a slightly edited form.

> The surface of any shell may be generated by the revolution about a fixed axis of a closed curve, which, remaining always geometrically similar to itself, increases its dimensions continually. [...] Let us imagine some characteristic point within this closed curve, such as its centre of gravity. Starting from a fixed origin, this characteristic point describes an equiangular spiral in space about a fixed axis (namely the axis of the shell), with or without a simultaneous movement of translation along the axis. The scale of the figure increases in geometrical progression while the angle of rotation increases in arithmetical, and the centre of similitude remains fixed. [...] The form of the generating curve is seldom open to easy mathematical expressions.

The construction of models derived from this description is presented below.

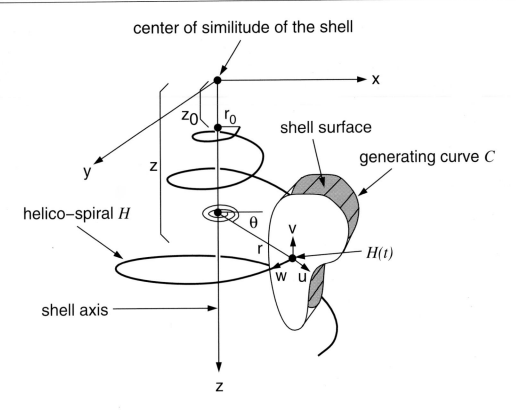

Figure 10.2. Construction of the shell surface

10.3 The helico-spiral

The modelling of a shell surface starts with the construction of a logarithmic (equiangular) helico-spiral $\mathcal{H}$ (Figure 10.2).

In a cylindrical coordinate system (shown in Figure 10.2 as embedded in the Cartesian xyz system) it has the parametric description (Coxeter, 1961):

$$\theta = t, \quad r = r_0 \xi_r{}^t, \quad z = z_0 \xi_z{}^t. \tag{10.1}$$

Parameter t ranges from 0 at the apex of the shell to t_{max} at the opening. The first two equations represent a logarithmic spiral lying in the plane $z = 0$. The third equation stretches the spiral along the z-axis, thus contributing the helical component to its shape.

Distances r and z are exponential functions of the parameter t, and usually have the same base, $\xi_r = \xi_z = \xi$. As a result, the generating helico-spiral is self-similar, with the center of similitude located at the origin of the coordinate system xyz. Given the initial values θ_0, r_0, and z_0, a sequence of points on the helico-spiral can be computed incrementally using the formulae:

$$\begin{aligned}
\theta_{i+1} &= t_i + \Delta t &&= \theta_i + \Delta\theta, \\
r_{i+1} &= r_0 \xi_r^{t_i} \xi_r^{\Delta t} &&= r_i \lambda_r, \\
z_{i+1} &= z_0 \xi_z^{t_i} \xi_z^{\Delta t} &&= z_i \lambda_z.
\end{aligned} \tag{10.2}$$

Figure 10.3. Variations in shell shape resulting from different parameter values characterising the helico-spiral. Leftmost: turbinate shell ($z_0 = 1.9$, $\lambda = 1.007$). Top row: patelliform shell ($z_0 = 0$, $\lambda = 1.34$) and tubular shell ($z_0 = 0.0$, $\lambda = 1.011$). Bottom row: spherical shell ($z_0 = 1.5$, $\lambda = 1.03$) and diskoid shell ($z_0 = 1.4$, $\lambda = 1.014$). Values of $\lambda = \lambda_r = \lambda_z$ correspond to $\Delta\theta = 10°$.

While the angle of rotation θ increases in arithmetic progression with the step $\Delta\theta$, the radius r forms a geometric progression with the scaling factor $\lambda_r = \xi_r^{\Delta t}$, and the vertical displacement z forms a geometric progression with the scaling factor $\lambda_z = \xi_z^{\Delta t}$. In many shells, parameters λ_r and λ_z are the same. Variations in shell shape due primarily to different parameter values characterising the helico-spiral are shown in Figure 10.3. They correspond closely to the shell types identified by Thompson (Thompson, 1961, page 192).

The parameters introduced in this section provide a convenient means for generating the logarithmic helico-spiral. Other mathematically equivalent families of parameters have been described in the literature (for example, see Cortie, 1989, Lovtrup and Lovtrup, 1988).

Figure 10.4. Variations in shell shape resulting from different generating curves. From left to right: turreted shell, two fusiform shells, and a conical shell.

10.4 The generating curve

The surface of the shell is determined by a generating curve $\mathscr{C}$, sweeping along the helico-spiral $\mathscr{H}$. The size of the curve $\mathscr{C}$ increases as it revolves around the shell axis. The shape of $\mathscr{C}$ determines the profile of the whorls and of the shell opening. In order to capture the variety and complexity of possible shapes, we construct the generating curves from one or more segments of Bézier curves (Foley *et al.*, 1990). This method makes it easy to design an almost unrestricted range of generative curves using an interactive three-dimensional modelling program. The impact of the generating curve on the shape of a shell is shown in Figures 10.4 and 10.5.

10.5 Incorporating the generating curve into the model

The generating curve $\mathscr{C}$ is specified in a local coordinate system uvw. Given a point $\mathscr{H}(t)$ of the helico-spiral, $\mathscr{C}$ is first scaled up by the factor $\xi_c{}^t$ with respect to the origin O of this system, then rotated and translated so that the point O matches $\mathscr{H}(t)$ (Figure 10.2). The axes uvw are used to orient the generating curve in space. The simplest approach is to rotate the system uvw so that the axes v and u become,

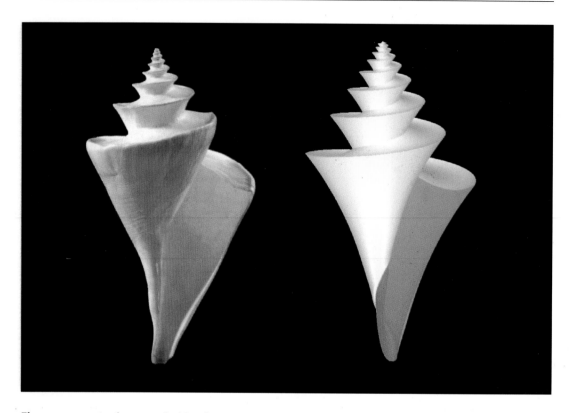

Figure 10.5. A photograph (Gordon, 1990, page 97) and model of *Thatcheria mirabilis* (Miraculous Thatcheria). The unusual shape of this shell results from the triangular generating curve. Photograph courtesy of the Natural History Museum, London, England.

respectively, parallel and perpendicular to the shell axis z. If the generating curve lies in the plane uv, the opening of the shell and the growth markings (such as the ribs on a shell surface) will be parallel to the shell axis. However, many shells exhibit approximately *orthoclinal* growth markings, which lie in planes normal to the helico-spiral $\mathcal{H}$ (Illert, 1989). This effect can be captured by orienting the axis w along the vector $\vec{e}_1$, tangent to the helico-spiral at the point $\mathcal{H}(t)$. The curve is fixed in space by aligning the axis u with the principal normal vector $\vec{e}_2$ of $\mathcal{H}$. The unit vectors $\vec{e}_1$ and $\vec{e}_2$ can be calculated as follows (Bronsvoort and Klok, 1985):

$$\vec{e}_1 = \frac{\vec{\mathcal{H}}'(t)}{\left|\vec{\mathcal{H}}'(t)\right|}, \quad \vec{e}_3 = \frac{\vec{e}_1 \times \vec{\mathcal{H}}''(t)}{\left|\vec{e}_1 \times \vec{\mathcal{H}}''(t)\right|}, \quad \vec{e}_2 = \vec{e}_3 \times \vec{e}_1. \tag{10.3}$$

Symbols $\vec{\mathcal{H}}'(t)$ and $\vec{\mathcal{H}}''(t)$ denote the first and second derivative of the position vector $\vec{\mathcal{H}}(t)$ of the point $\mathcal{H}(t)$, taken with respect to the parameter t. Vectors $\vec{e}_1$, $\vec{e}_2$ and $\vec{e}_3$ define a local orthogonal coordinate system called the *Frenet frame*. It is considered a good reference system for specifying orientation, because it does not

Figure 10.6. A photograph (Gordon, 1990, page 47) and two models of *Epitonium scalare* (Precious Wentletrap). Photograph courtesy of Ken Lucas, Biological Photo Service, Moss Beach, California.

depend on the parametrisation of the helico-spiral $\mathcal{H}$ or on the coordinate system in which it is expressed (do Carmo, 1976). The Frenet frame is not defined at points with zero curvature, but a helico-spiral has no such points ($\mathcal{H}''(t)$ is never equal to zero). The impact of the orientation of the generating curve on the shell shape is illustrated in Figure 10.6.

The opening of the real shell and the ribs on its surface lie in planes normal to the helico-spiral. This is properly captured in the model in the center of the figure, which uses Frenet frames to orient the generating curve. The model on the right incorrectly aligns the generating curve with the shell axis.

In general, the generating curve need not be aligned either with the shell axis or with the Frenet frame (Cortie, 1989, Illert, 1989). In the case of non-planar generating curves, it is even difficult to define what "alignment" would mean. It is therefore convenient to be able to adjust the orientation of the generating curve with respect to the reference coordinate system. We accomplish this by allowing the user to specify a rotation of the system *uvw* with respect to each of the axes $\vec{e}_1$, $\vec{e}_2$, and $\vec{e}_3$.

Although, in a mathematical sense, the surface of a shell is completely defined by the generating curve $\mathcal{C}$ sweeping along the helico-spiral $\mathcal{H}$, we represent it as a polygon mesh for image synthesis purposes. The mesh is constructed by specifying $n + 1$ points on the generating curve (including the endpoints), and connecting corresponding points at consecutive positions of the generating curve. Such a polygonal representation can be easily rendered using standard computer graphics techniques (Foley *et al.*, 1990).

10.6   Modeling the sculpture on shell surfaces

Many shells have a sculptured surface. Common forms of sculpturing include ribs more or less parallel to the generating curve or to the direction of growth. Both types of ribs can be easily reproduced by displacing the vertices of the polygon mesh in a direction normal to the shell surface. Ribs parallel to the generating curve result from a periodical variation of the value of the displacement d according to the position of the generating curve along the helico-spiral $\mathcal{H}$. The amplitude is proportional to the current size of the generating curve. A striking example of such ribs can be observed in Precious Wentletrap (Figure 10.6). A shell with more gentle corrugation is shown in Figure 10.7.

In the case of ribs parallel to the direction of growth, the displacement d varies periodically along the generating curve. As previously, the amplitude of these variations is proportional to the current size of the curve. Examples of this type of sculpturing are shown in Figures 10.8, 10.9 and 10.10.

More dramatic departures from the "basic" shell shape can be reproduced by combining several generating curves in the same model. For example, the knobs on the surface of a triton shell (Figure 10.11) were generated by interpolating between two periodically exchanged generating curves: one smooth and one undulating. A similar method was used to model the trophon shell in Figure 10.1. In this case, one of the curves had cusps, which produced the spikes on the shell surface.

Three generating curves can be found in the model of the pelican shell shown in Figure 10.12. Two alternating curves were used to generate most of the shell body, as in the triton shell. The third curve, with large cusps, captured the shape of the shell opening.

Figure 10.7. A corrugated shell

Figure 10.8. Clam shells provide a simple example of surface sculpturing with ribs orthogonal to the generating curve.

Figure 10.9. A photograph (see Sabelli, 1979) and model of *Rapa rapa* (Papery Rapa) showing surface sculpturing with ribs orthogonal to the generating curve. The shape of the ribs in the model is captured by a sine function uniformly spaced along the edge of the shell.

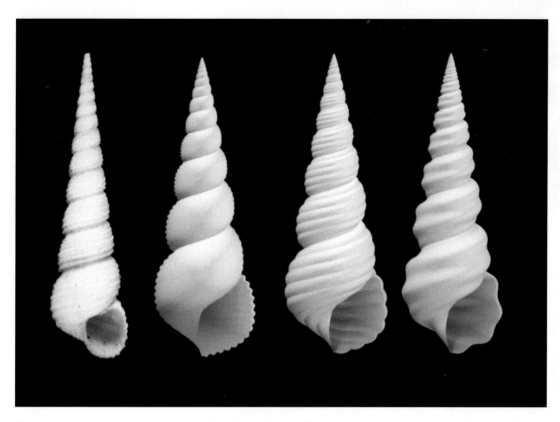

Figure 10.10. Surface sculpturing with ribs orthogonal to the generating curve. A photograph (see Sabelli, 1979) and three models of *Turritella nivea* illustrate the effect of decreasing the frequency of the modulating function.

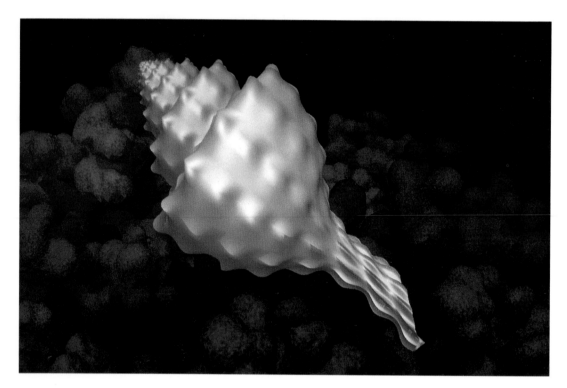

Figure 10.11. A triton shell. The corals in the background were modeled by Jaap Kaandorp, using a method described in his book (Kaandorp, 1994; chapter 5).

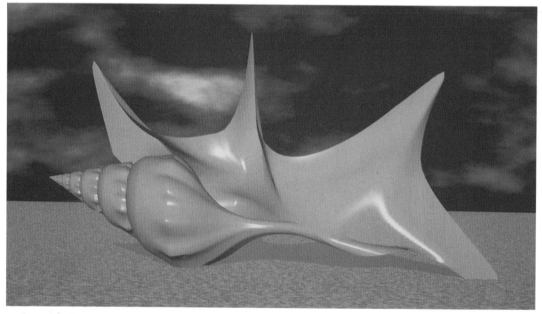

Figure 10.12. A pelican shell

10.7 Shells with patterns

As discussed in the previous chapters, patterns in shells result from the deposition of pigmented material at the shell margin. In this section, we illustrate this process using synthetic images of selected shells.

The basic technique for incorporating patterns into three-dimensional shell models is depicted in Figure 10.13. Instead of developing a pattern along a horizontal line moving downwards (left), we solve differential equations representing pigment deposition along the edge of the shell (right). The pattern unfolds on the shell surface as the shell grows[1].

---

[1]Technical details of the mapping of patterns to polygon meshes are discussed in Fowler *et al.*, (1992)

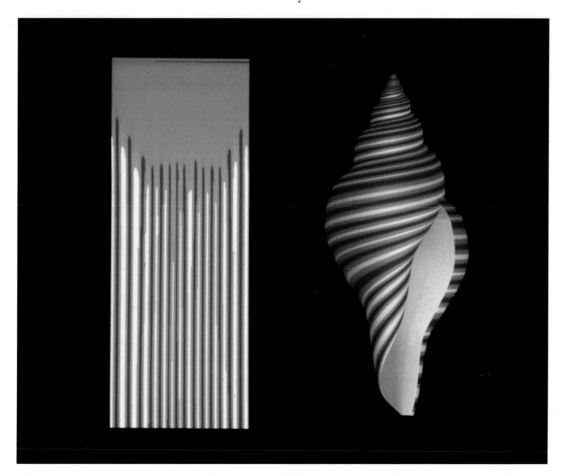

Figure 10.13. A stable pattern of stripes generated by the activator-substrate model (Equation 2.4)

Figure 10.14. A photograph and model of *Amoria ellioti*

The particular patterns shown in Figure 10.13 were generated using the activator-substrate model introduced in Chapter 2 (Equation 2.4). The parameter values were chosen to produce a stable distribution of pigmentation along the growing edge, yielding stripes parallel to the direction of shell growth.

Another pattern generated using the activator-substrate model, stripes perpendicular to the direction of growth, can be found in *Amoria ellioti* shown in Figure 10.14. As described in Chapter 3, the periodic character of this pattern is a manifestation of the oscillations of the activator concentration over time, which occur when the basic substrate production b_b is not sufficient to sustain activator removal at a constant rate r_a ($b_b < r_a$ in Equation 2.4).

Figure 10.15 shows a photograph and a model of *Amoria undulata*. Its pattern is a modification of that found in *Amoria ellioti*, with undulations superimposed on lines parallel to the growing edge. As explained in Sections 4.1 and 4.2, the activator-substrate process is regulated in this case by an additional pattern that modulates the substrate production b_b according to a periodic (sine) function of

Figure 10.15. A photograph (see Sabelli, 1979) and model of *Amoria undulata* (Waved Volute)

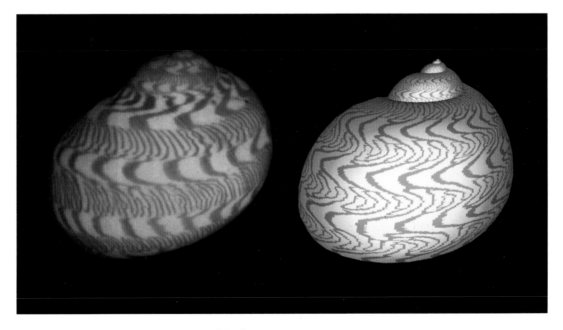

Figure 10.16. A photograph and model of *Natica euzona*

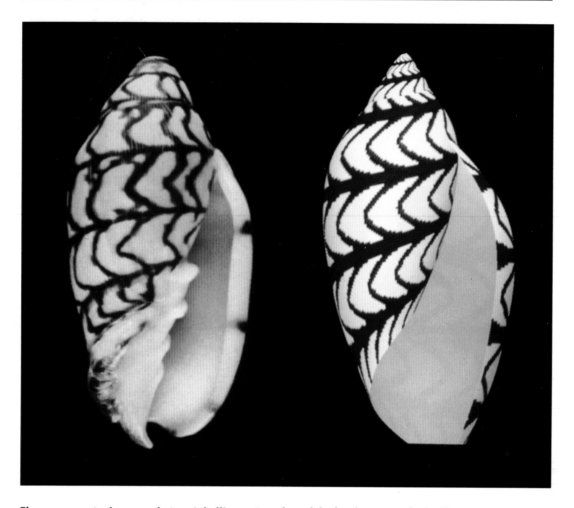

Figure 10.17. A photograph (see Sabelli, 1979) and model of *Volutoconus bednalli* (Bednall's Volute)

the cell position, $b_b = b_b(x)$. Undulations occur because oscillations are faster in regions with higher b_b values than in regions with lower values. The coherence of the lines is maintained by the diffusion of the activator.

As the modulation of the substrate production increases, the undulations become even more pronounced, as observed in *Natica euzona* (Figure 10.16). Regions of high and low frequency of oscillations can be distinguished along the growing edge of the shell. Diffusion of the activator is insufficient to maintain the coherence of the lines, which merge and form blind endings as described in Section 4.1 (Figures 4.3 and 4.4).

Volutoconus bednalli (Figure 10.17) illustrates an extreme case of the same mechanism (see Section 4.7). The function $b_b(x)$, which describes the production of the substrate in Equation 2.4, exceeds the decay constant r_a for some locations x along the growing edge of the shell. This creates areas of permanently high activator concentration and yields a pattern of lines parallel to the direction of growth. Oscillations occur between these lines, producing arcs roughly parallel to the growing edge.

Figure 10.18. A photograph (see Sabelli, 1979) and model of *Oliva porphyria*

The formation of branches is an essential feature of the pigmentation pattern of *Oliva porphyria*, presented in Figure 10.18. Oblique lines represent waves of activator concentration, traveling along the growing edge. Colliding waves extinguish each other. New waves appear when the activated point of one wave spontaneously initiates another wave traveling in the opposite direction. Observations of the shell indicate that the number of traveling waves is approximately constant over time. This suggests a global control mechanism, or "hormone" that monitors the total amount of activator in the system, and initiates new waves when its concentration becomes too low. This mechanism has been modeled using a modified activator-inhibitor system (Equations 2.1 and 6.1), as described in detail in Section 6.2.

The model of *Conus marmoreus*, shown in Figure 10.19, is similar to that of *Oliva*. As described in Chapter 7, in this case the pigment producing process is controlled by another reaction-diffusion process, rather than a hormone.

The three-dimensional shell models collected in this chapter (Figure 10.20) present a small fraction of patterns described in this book. Other patterns can be incorporated into the models in a similar fashion.

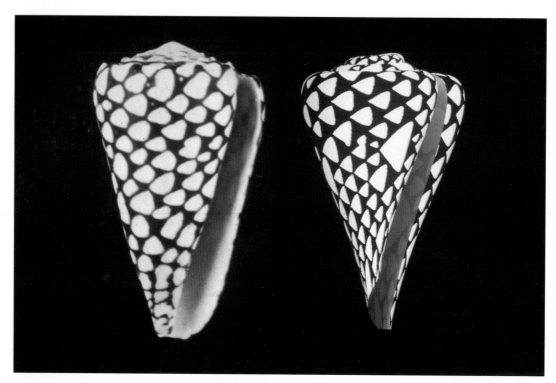

Figure 10.19. A photograph (see Sabelli, 1979) and model of *Conus marmoreus* (Marble Cone)

Figure 10.20. A virtual museum of shells

This chapter contains edited parts of the paper *Modeling seashells* (Fowler, *et al.*, 1992), which is used with the permission of the Association for Computer Machinery. The reported research was sponsored by operating and equipment grants from the Natural Sciences and Engineering Research Council of Canada, and by a graduate scholarship from the University of Regina. We are indebted to Pat Hanrahan for providing Deborah with access to his research facilities at Princeton. Images 10.1, 10.8, 10.7 10.11 10.12 10.16 10.20 were rendered using the ray-tracer rayshade by Craig Kolb. Photographs of real shells included in Figures 10.9, 10.10, 10.15, 10.17, 10.18, and 10.19 are reproduced with the kind permission of Giuseppe Mazza.

For those in hurry: a quick start to the computer program

The program should run on any IBM-compatible computer. A floating point processor is highly recommended. At least 550 KB of free memory, a 486-DX processor and a VGA colour display would be ideal.

The program may be started from diskette but it is usually more convenient to install it on the hard disk. Make a new directory (for instance, SP). Go to that directory and copy in the contents of the diskette, maintaining the subdirectory structure. For example, use the following command:

```
xcopy a:*.*/S  (or xcopy b:*.*/S )
```

From the DOS-level prompt, start the program by typing

```
SP <RETURN>
```

A list of the most frequently used commands is given on the initial screen. Other commands and parameter changes can be selected from a menu by pressing F1.

Most figure captions in this book contain commands to reproduce the corresponding simulations. Such a command consist of the letter S (Simulation) followed the figure number. For instance, S61 or S6.1 will produce the simulation shown in Figure 6.1. Equations not explicitly given in the book may be found by reading in the parameters of the corresponding simulation (for instance, r61) and using the PE (**PrintEquation**) command.

Simulations that require more complex input from the keyboard have been simplified using "GUIDED TOURS". In this case, the input is read from a file step by step by pressing the <RETURN> key. For instance, the command GT32 leads to the simulations shown in Figure 3.2.

If a monochrome screen is used, it is recommended that only the distribution of pigment-producing substance a be displayed. The commands required to do this are
DW <RETURN>
a <RETURN>

Chapter 11

The computer program

11.1 Introductory remarks

The program supplied is a modified version of one of my own working programs. Originally written in FORTRAN, it has been translated into BASIC and compiled with Microsoft Professional Basic PDS 7.1 and Power Basic 3.0. It should also be possible to recompile it using Microsoft Visual Basic for DOS. Do not expect the program to be as perfect as a commercial product. Consider it as an extension of the book and a tool to discover more intuitive inroads into the complex events connected with non-linear interactions. I am sure that the program is not free from errors and I cannot exclude the fact that it contains awkward parts left over from earlier versions. However, since the source code is provided, it should be possible to make corrections or improvements if desired. Conversion from BASIC to other programming languages should not be too difficult.

The program does not have a fancy interface and it may require some habituation. However, changing parameters and inputting new commands requires only a few key strokes so that rapid and convenient operation should be possible. Certainly the best way to become familiar with the commands is to exercise them. For those who do not want to delve too deeply into the program but would like to reproduce the simulations shown in this book, it may be encouraging to know that the very simple commands given in the figures captions are sufficient.

11.2 Using the program

The installation is described on page 183. The program will work in any standard DOS environment. The use of a floating point processor is highly recommended, otherwise the simulations will be boringly slow. A 486-DX processor and at least 550KB of free memory would be ideal.

To start the program, type SP. The first screen is subdivided into two parts (Figure 11.1). The upper part contains the commands most frequently used as well as a list of the function keys. The lower part contains a list of parameters read from the prerecorded parameter file SP1.PRM. The arrow $(->)$ at the bottom

```
--------        PROGRAM FOR THE SIMULATION OF SEA SHELL PATTERNS    ------
Commands: S = Start of simulation, Q = Quit the program;
          C = Continuation,        N = Continuation with a New screen
Simulations can be interrupted anytime using <ESC>
a <RETURN> displays parameter list again, a second <RETURN> the initial screen


  FUNCTION KEYS:
  F1:   HELP Menu                 F6 Print equation on screen or lpt1:
  F2:   Parameter for a chapter   F7 Print parameters on lpt1:
  F3:   Read next parameter set   F8 Store parameters in file SP0.prm
  F4:   Show distributions        F9 Read parameters temporary stored with F8
  F5:   Print distributions       F10 Simulation with next parameter set


  R   = Read = get a list of the pre-recorded parameter files and select one
  Rxx = Read the pre-recorded parameter file xx, xx=0-999
  Sxx = Start directly with the pre-recorded parameter file xx, xx=0-999
  Wxx = Write present parameters as file SPxx.PRM




  Progr.: SPMS.BAS; read from file c:\sp\param\SP1.prm
                   Stable periodic pattern in space
    45-KT    12-KP    1-KX   20-KY    1-KD    1-KI   21-KE   3-KR   2-KN   0-KG
    2-K1     2-K2    0-K3    0-K4          16.0000-DX  0.0000-DY 12.0000-DZ  ab-DW
    0.0100-DA  0.0500-RA  0.0100-BA  0.0000-SA  0.0000-CA  1.5000-AA  1.5000-GA
    0.4000-DB  0.0000-RB  0.0000-BB  0.0000-SB  0.0000-CB  1.5000-AB  1.5000-GB

  ->█
```

Figure 11.1. The initial screen with basic commands and parameters read from the file SP1.PRM

is the prompt, indicating that the program is waiting for input. The input can be either a command or the name of a parameter. Either upper or lower case may be used. A calculation can be interrupted at any time by pressing the <ESC> key. Q (quit) terminates the program and returns to the DOS prompt. As mentioned, most important commands are listed on the screen. A more complete list can be accessed by pressing F1. Commands may be selected and parameters changed using this menu.

Each parameter in the list has a name. The names are the same as those used in the equations. The value of a parameter may be changed by entering its name. For example, type DA or da to change the value of the diffusion constant of the activator (D_a in the equations). The computer will first display the current value. You can either type a new value or <ESC> to leave the parameter unchanged. Pressing <RETURN> will set the parameter to zero. Table 11.1 provides a list of parameters.

The first 14 parameters are integer numbers and are used to control the program flow. They include number of the iterations, the number of iterations between updating the display, type of display, the type of equation or initial conditions (see

Table 11.2). The rule that parameters starting with the letters H through N are integers has been maintained from the original FORTRAN program.

After each simulation, only the prompt line is redisplayed at the bottom of the screen, leaving the graphic output undisturbed. At this point, parameters can be changed or commands can be given as described above. Pressing the <RETURN> key lists the complete set of parameters. A second <RETURN> redisplays the initial screen with the list of commands.

Prerecorded parameter files can be read with the command Rx, where x is a number between 1 and 999 optionally followed by a letter. For example, R1 reads the file SP1.prm that was used at the start of the program. Typing R without a number will produce a list of the available prerecorded parameter files and the headlines of the corresponding simulations, with the option to select one of the files. The number x usually corresponds to the figure number. For instance, the command R23a will read the parameters required for simulating Figure 2.3a (you can, but need not, type the dot). With the full parameter list a message is displayed if the parameters have been modified. The original parameters can be re-read using the command RR. The command Sx may be used to directly start a simulation. These commands are given at the end of the figure captions (for instance, S22). To save a modified set of parameters use the command is Wx (w = write; again x is the file number, x = 1...999); WW saves a parameter set using the current file name. Subpictures such as Wxa are also legal. No warning is given when existing files are overwritten.

Simulations that require more complex input from the keyboard have been simplified using "GUIDED TOURS". The GT command offers a list of the available tours from which to choose. For example the command GT32 will produce the simulations shown in Figure 3.2. With a GUIDED TOUR, the data that is normally input from the keyboard is read from a file (in this example, the file name is SP32.GT). Comments are displayed to explain what is happening. In order to follow the input more easily, a time delay can be selected using the command igtdelay (in tenths of a second). If igtdelay is greater than 50, a <RETURN> is required after each input from the file.

11.3 Implementation of the interactions

After setting the initial conditions, the program calculates the concentration changes over a given time interval. These changes are added to the existing concentrations and the next iteration is started, and so on. After a number of iterations (as defined by the parameter KP), the program plots the new concentrations. This will repeat KT times such that KT * KP iterations are calculated in all.

The adaptation of the equations into a form convenient for computer simulations is illustrated using equation 2.1a:

$$\frac{\partial a}{\partial t} = s \left(\frac{a^2}{b} + b_a \right) - r_a a + D_a \frac{\partial^2 a}{\partial x^2}$$

Rather than considering a homogeneous space in which concentrations vary continuously, the space is subdivided into individual cells i, $i = 1...n$. The concentrations of substances a, b, ... are stored in the array $a(il,i)$; where $il = 1, 2, ...$ is the number of substance ($il = 1$ corresponds to substance a, $il = 2$ to substance b and so on) and i the position of the cell. Before the actual computation is made for a particular cell and time step, the actual concentrations of the substances $a, b, c, ...$ are stored in the floating point numbers a, b, c, ... As a rule, the constant term in the production rate, s (source density), is set to the decay rate of a in order to obtain absolute concentrations around unity. This value is modulated by random fluctuations (parameter KR = $\pm$ fluctuations in percent). The local source density is stored in the array $a(i,0)$. Random fluctuations remain unchanged during a simulation. In the calculations of an individual cell the local value of the source density $a(i,0)$ is stored in the variable s .

The loss or gain by diffusion also depends on the difference in concentration between two neighbouring cells. Thus, the change in activator concentration at time t is

$$da_{i,t} = s * (a_{i,t}^2 / b_{i,t} + ba) - ra * a_{i,t} + DA * ((a_{i-1,t} - a_{i,t}) + (a_{i+1,t} - a_{i,t}))$$

Adding this change to the existing concentration of a at the time t leads to a new concentration of a in the cell i at the time $t + 1$

$$a_{i,t+1} = a_{i,t} + da_{i,t}$$

Communication between cells using diffusion requires special conditions at boundaries where cells have only a single neighbour. In all calculations it is assumed that the boundaries are closed, i.e., no loss or gain takes place through the boundary. This is achieved by the assumption of virtual leftmost and rightmost cells that have the same concentration as the cells at the boundary proper. Since no net exchange takes place between cells with identical concentrations, the no-flux boundary condition is satisfied.

Since the program contains many types of interaction, the common elements of all calculations for decay and diffusion are coded separately from the interaction proper. Factors that do not change during the simulation are also precalculated in order to speed up the calculations. With each iteration new concentrations that result only from decay and diffusion are calculated first since they are common to all interactions. In the program, these values are termed olddecaydiffA for substance *a*, etc. Thus, for the example given above, the new concentration of *a* in the cell *i* would be

```
a(1,i) = olddecaydiffA + s * (a * a / b + ba)
```

Due to its similarity with the original equation, this type of notation should be understandable even by non-computer specialists. The command PE (PrintEquation) provides the code for the interaction used in the actual simulation. The program reads lines of source code according to the equation-selecting parameter KE and includes the option to print these lines. (The source program SP.BAS must be in the actual directory). The following example shows the code for the activator-depleted substrate model (KE = 24, equation 2.4) as printed by this command.

```
CASE 24 '- activator-depletion mechanism: ----
     ' a activator, b substrate,
     ' a(8,i) may contain a stable pattern (normalized to 1)
     aq = s * b * (a * a / (1 + sa * a * a) + ba)
     a(1, i) = olddecaydiffA + aq
     a(2, i) = olddecaydiffB - aq + bb * a(8, i)
```

11.4 Numerical instabilities that may cause errors

Concentration changes are calculated for a finite time interval. After each interval the changes are added to existing concentrations to obtain the new values. However, in reality the concentrations would be changing continuously over time. Therefore, to obtain a close approximation of the real situation very short time steps should be used. This may result in a prohibitive number of calculations if the total time frame is fairly long. A compromise between speed and precision must be made.

There are also some instances in which step-by-step calculations may become numerically instable. This may be caused by diffusion rates which are too high. For example, imagine a chain of cells in which only a single cell has a high concentration. With a numerical diffusion constant of 1 this cell will obtain a negative concentration (of the same absolute value) in the next iteration while both neighbouring cells will have the same concentration as the original cell. This is, of course, senseless since concentration differences should only smooth out by diffusion, not enlarge. Therefore, to avoid these instabilities the numerical value of diffusion constants should be smaller than 0.4. The program will display a message and refuse to accept larger diffusion rates.

Another possible source of numerical instability exists in simulations of the depletion mechanism. During the autocatalytic burst, the extrapolation of a given substrate removal rate may lead to negative substrate concentrations, which, again, are not possible. This causes numerical instabilities since the sign change means the substrate is no longer being removed but is increasing. In this case, smaller time steps may cure the problem, requiring smaller constants for production, decay and diffusion. A simpler way is to introduce a small saturation term (s_a in equation 2.4a) to limit the autocatalytic burst. Usually this avoids the instability without requiring more computer time and does not significantly change the outcoming pattern. In critical interactions negative concentrations are explicitly prevented by the program.

11.5 Compilers and versions

The program comes in two versions: SPMS.BAS, SPMS.EXE for the Microsoft BASIC family of compilers (Quick, Professional or Visual BASIC for DOS); and SP.BAS, SP.EXE for Power-BASIC 3.0. The program codes are identical except for some language-specific elements. Both versions have their advantages. One problem with the MS-BASIC family is the conflict between the EMM386 - driver with the floating point processor. If this driver is active (i.e., declared in CON-FIG.SYS) the computations are slower by approximately a factor of 3. This driver is usually installed on machines using WINDOWS. On the other hand, the Microsoft-compiled versions are usually very stable. The POWER-BASIC version does not have this conflict and is, therefore, recommended for standard use. Both versions use identical subroutines (plib.bas).

When a HERCULES monochrome graphic card is used with the Microsoft version, MSHERC, a standard driver delivered with Microsoft's DOS 5.0 and higher must be loaded.

If the source code is changed and the program must be recompiled, more computer memory will be needed. Depending on the configuration it may be necessary to compile the Power-BASIC version with the Command Line Compiler, rather than the Integrated Development Environment. The corresponding command is PBC SP.

11.6 Parameters used in the program

The following tables describe the parameters used in the program in more detail. These tables are also part of the menu system that can be called in with F1. The parameters are divided into several categories:

- floating point variables (rates of diffusion, life times, etc.)
- parameters for program control (number of iterations, type of interaction etc.)
- initial conditions
- types of display
- general display parameters

Floating point variables

Parameter names consist of two letters. The first indicates the function (D=Diffusion, R=Removal...), the second the substance A, B, C... to which it applies.

DA, DB, DC, ...: Diffusion constants for A, B, ... Diffusion constants larger than 0.4 lead to numerical instabilities and are correspondingly corrected. A message is displayed.

RA, RB, RC, ...: Removal (decay) rates

BA, BB, BC, ...: Basic (activator independent) production

SA, SB, SC, ...: Saturation and Michaelis-Menten constants (and other use)

CA, CB, CC, ...: Coupling between several systems

AA, AB, AC, ...: Initial concentrations at particular positions

GA, GB, GC, ...: General initial concentrations

The following parameters play a special role:

CA Usually the general production strength (source density s) is set to the decay rate r_a. If CA is set to a value > 0, s is set to this value rather than to the decay rate. Thus, with CA=0, a change in the decay rate automatically leads to a corresponding change in the production rate.

DX pixels per cell in the graphic display

DY used for some initial conditions, see Table 11.3

DW "DisplayWhat" is a string variable and determines the substances to be plotted (see Table 11.5).

Table 11.1. Floating point variables

List of integer parameters to control program flow

KT Total number of plots per simulation

KP Number of iterations (time steps) between two plots. The total number of iterations is KT * KP

KX Left-most cell of a field, usually =1 (or X-field size in two-dimensional simulations)

KY Right-most cell of a field (or Y-field size in two-dimensional simulations)

KD Type of display, see Table 11.4

KI Initial condition, see Table 11.3

KE Equation to be used, usually the equation number in the text, e.g., 21 for equation 2.1

KR Extent of random fluctuation superimposed, in percent; stored in the array elements a(0,i), substance s (source density) in the display

KN Number of substances, e.g. 2 if substances a and b are involved

KG Growth, 0 if no growth takes place. Insertion takes place after KG plots are displayed, i.e after KG * KP iterations. The location of new cells is determined by K1

K1 Mode of growth if KG>0 (for other uses of K1-K4 see Table 11.3)

    0 Insertion at both terminal ends (at the left side only if space is available, i.e. if KX >1.

    1 Two insertions at random positions, one in the right and one in the left half of the field (see Figure 2.6)

    2 Insertion in the centre

    3 Insertion in the right third of the field

    4 Insertion at a random position

K2 see Table 11.3

K3 After K3 plots, manipulations or parameter changes may be evoked. The type of manipulation is determined by K4. No manipulation takes place if K3 = 0.

K4 Type of manipulation

    1 Basic a (activator) production (BA) is changed to DZ

    2 Decay rate of b (RB) is changed to DZ

    3 Basic inhibitor / substrate production (BB) is changed to DZ

    4 Decay rate of the activator (RA) is changed to DZ

    5 a (activator) concentration is changed in a fraction of the field

    6 a concentration is changed across the entire field by a factor DZ

    7 Production and decay rate of c (RC) is changed to DZ

    8 Saturation of the activator SA is changed to DZ

    9 Coupling constant CB is set to DZ

    10 c-concentration is changed in a fraction of the field

    11 Both activator and inhibitor diffusion rate is set to zero

Table 11.2. List of integer parameters to control program flow

Initial conditions (parameter KI)

The first possibilities (KI=1...6) set the *a*, *c* and *e* concentrations of specific cells to the concentrations given by parameters AA, CA and EA respectively.

1 Left-most cell

2 Left-most and right-most cell

3 Central cell

4 Cell numbers are input from the keybord (0=end of selection)

5 Specific cells (10, 22, 55, 75, 115, 180, 195, 240, 310, ...)

6 Cells at random positions within a 10-50 cell distance

The following choices are available for the generation of stable prepatterns that act on substrate production. The patterns are stored in the array a(8,i). This mode is used in the simulations in chapter 4.

7 A sinusoidal prepattern is generated, maximum value = 1 (see Figure 4.9). Other variable settings required:

DY Degree of modulation, DY=1 causes a ratio of 1 to 0.5
K1 Repeat length of the pattern in cells
K2 Position of the maximum (phase)

8 Source density *s* has an exponential distribution, exponent given by AB.

9 A more step-like prepattern is generated (see Figure 4.2). Other variable settings required:

DY Degree of modulation, DY=1 causes a ratio of 1 to 0.5
K1 Repeat length of the pattern in cells
K2 Start of the HIGH-cells
K4 End of the HIGH-cells
DZ Degree of smoothing, 1-80 is reasonable
AB a superimposed linear gradient

10 Left-most cell can have a different rate of substrate production. The factor is given in AB. This cell may act as a pace-maker region (see Figure 3.8)

11 Left-most and right-most cell may act as pace-maker regions (see Figure 3.9)

Table 11.3. Initial conditions (parameter KI)

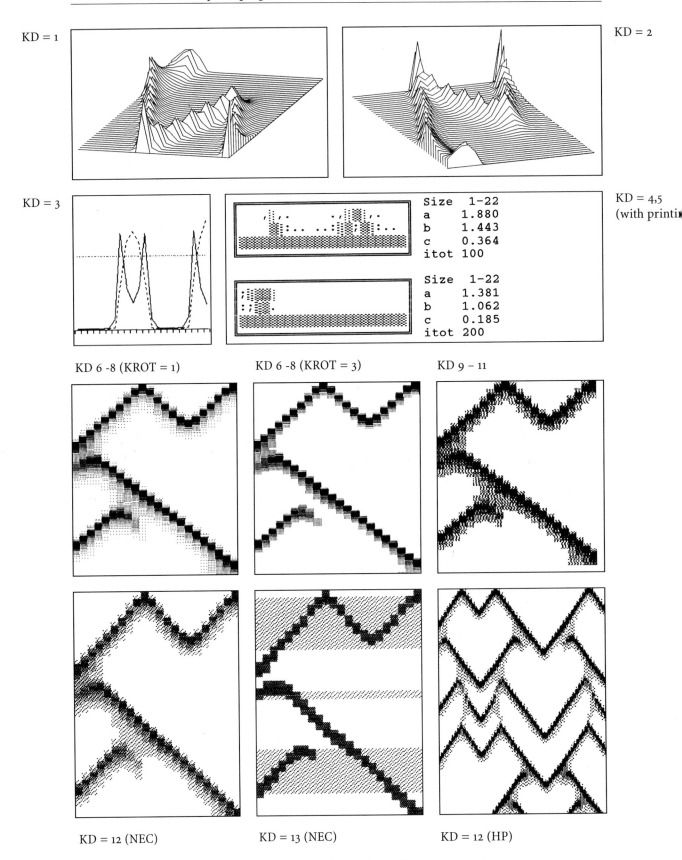

Figure 11.2. Sample modes of display, for details see the Table 11.4 (opposite page)

Types of plots (parameter KD)

Examples are given in Figures 11.2 and Figure 11.3

1 Three-dimensional XT-plot, the past is in the foreground
2 Three-dimensional XT-plot, the past is in the background
3 Single curves (width can be controlled by the command IWI)
4 Schematic plot with symbols and maximum concentrations
5 Same as 4, with output to the printer
6 Concentration plotted in false colour. This mode produces a fine gray scale for back-and-white screen prints. Use only when plotting a single array, e.g. DW = a, or side by side using different X-coordinates. A lower threshold level can be introduced by setting KROT > 1
7 False colour, landscape orientation
8 False colour + plot of stable prepattern (array a(8,i))
9 Pixels proportional to concentration. Several arrays may be plotted on top of each other if they have the same X-position, or side by side if the X-positions are different
10 Pixels proportional to concentration, landscape orientation. Appropriate for recording development over a long period of time (KT up to 630).
11 Pixels proportional to concentration and stable pattern (array a(8,i)), portrait orientation. See Figure 4.5
12 as 9. After the simulation a schematic plot of the activator on a HP or NEC-Printer is possible. (When using a NEC printer give first once command NEC). No print-screen facility is required. The NEC printer will plot up to 100 cells; the HP printer, up to 250 cells.
13 as 12, the schematic plot includes the substances a and c in different grey levels if their concentrations are above thresholds.
14 Plot with a single threshold
15 Circular plot used for the Nautilus shell (see Figure 4.8)
16 Plot around a circular arrangement of cells (not used here)
17 Plot in different brownish colours (MS-version, see Figure 9.8)
19 Similar to KD=3 but areas are filled (see Figure 6.1c)
20 Plot of concentrations in a single cell as a function of time; use with KY=1 (see Figures 3.2 and 3.3)

Table 11.4. Types of plots (parameter KD)

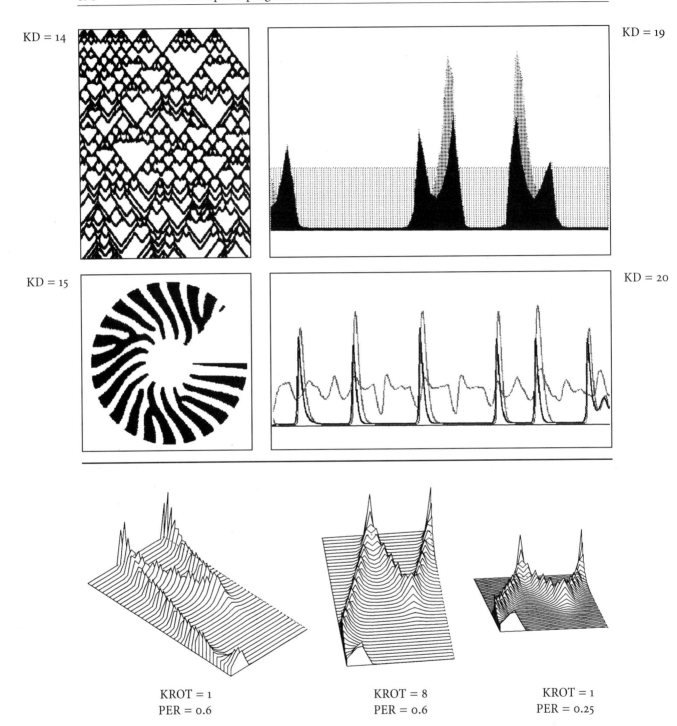

Figure 11.3. Top: Sample modes of display, continued from Figure 11.2 (see Table 11.4). Bottom: Examples of the influence of graphic display parameters. For details, see Table 11.5

Variables used for the graphic display

Examples are given in the lower part of Figure 11.3

DX Unit length of a cell in pixels, 2-6 is reasonable

PER Perspective in 3D-display mode. 0 is a front view, 1 is more from above, 0.2 - 0.6 is reasonable

KROT Side view in 3D-display mode, 1 is a more lateral view, higher numbers produce a more frontal view, but require more time and memory. If memory is exhausted, the plot will terminate. This may happen if several large arrays are plotted. For false-colour plots (KD=6-8), KROT sets a lower threshold

BCOL Background and foreground colour

DW "DisplayWhat" is a string and determines the substances to be plotted. For example, **ab** displays substances *a* and *b* [activator and inhibitor, array elements a(1,i) and a(2,i)]. The sequence determines the plotting order. If DW=ab, the *a* distribution will be plotted on top of *b*, if DW=ba they will be plotted in the reverse order

The following parameters allow input of display characteristics for each substance chosen with DW (DisplayWhat). The input may be terminated by **q**; input of <ESC> leaves a particular parameter unchanged, and causes the request of the next. Beginning a parameter value with the letter **g** allows global definition, e.g. **X** <RETURN> **g50** <RETURN> sets all X-coordinates to 50 causing all distributions to be plotted on top of each other.

X X-position of the plot (in pixel-units, 640 * 480)

Y Y-position of the plot (in pixel-units)

F normalization factor of the plot (1 is reasonable if concentrations are around 1)

ICOL colour of the plot (0-15)

Either an EGA or VGA screen may be used. The EGA screen has a lower resolution but, since two screen pages are available, plots with frequent screen refreshes can be shown without flickering. Good for display modes KD=3 and KD=19 (see simulation S22).

EGA selects EGA screen

VGA selects VGA screen

Table 11.5. Variables used for the graphic display

References

Bär, M., Eiswirth, M., Rotermund, H.H. and Ertl, G. (1992). Solitary-wave phenomena in an excitable surface-reaction. *Phys. Rev. Lett.* 69, 945-948.

Bronsvoort, W. and Klok, F. (1985). Ray tracing generalized cylinders. *ACM Transactions on Graphics* 4, 291-303 .

Cook, T.A. (1914). Curves of Life. Constable and Company, London, reprinted 1979 by Dover Publications, New York.

Cortie, M.B. (1989). Models for the mollusc shell shape. *South African Journal of Science* 85, 454-460.

Coxeter, H.S.M. (1961). Introduction to Geometry. J. Wiley and Sons, New York.

Crick, F. (1970). Diffusion in embryogenesis. *Nature* 225, 420-422.

do Carmo, M. (1976). Differential Geometry of Curves and Surfaces. Prentice Hall, Englewood Cliffs

Ermentrout, B., Campbell, J. and Oster, G. (1986). A model for shell patterns based on neural activity. *The Veliger* 28, 369-338.

Feininger, A. and Emerson, W. K. (1972). Shells. The Viking Press.

Foley, J.D., van Dam, A., Feiner, S. and Hughes, J. (1990). Computer graphics: Principles and practice. Addison-Wesley, Reading.

Fowler, D.R., Meinhardt, H. and Prusinkiewicz, P. (1992). Modeling seashells. Proceedings of SIGGRAPH '92. In: *Computer Graphics* 26, 379-387.

Gierer, A. (1977). Biological features and physical concepts of pattern formation exemplified by hydra. *Curr. Top. Dev. Biol.* 11, 17-59.

Gierer, A. (1981). Generation of biological patterns and form: Some physical, mathematical, and logical aspects. *Prog. Biophys. molec. Biol.* 37, 1-47.

Gierer, A. and Meinhardt, H. (1972). A theory of biological pattern formation. *Kybernetik* 12, 30-39.

Glass, L. and Mackey, M.C. (1988). From clocks to chaos. Princeton University Press.

Gordon, N.R. (1990). Seashells: A Photographic Celebration. Friedman group, New York.

Granero, M.I., Porati, A. and Zanacca, D. (1977). A bifurcation analysis of pattern formation in a diffusion governed morphogenetic field. *J. Math. Biology* 4, 21-27.

Grüneberg, H. (1976). Population studies on a polymorphic prosobranch snail (*Clithon (Pictoneritina) oualaniensis* Lesson). *Phil.Transctions Royal Soc. London B* 275, 385-437.

Herman, G.T. and Liu, W.H. (1973). The daughter of Celia, the french flag and the firing quad: Progress report on a cellular linear iterative-array simulator. *Simulation* 21, 33-41.

Illert, C. (1989). Formulation and solution of the classical seashell problem. *Il Nuovo Cimento* 11 D, 761-780.

Kaandorp, J. (1994). Fractal modelling: Growth and form in biology. Springer-Verlag, Heidelberg

Kawaguchi, Y. (1982). A morphological study of the form of nature. *Computer Graphics* 16, 223-232.

Lindsday, D.T. (1982). A new programmatic basis for shell pigment patterns in the bivalue mollusc Lioconcha castrensis (L.). *Differentiation* 21, 32-36.

Lovtrup, S. and Lovtrup, M. (1988). The morphogenesis of molluscan shells: A mathematical account using biological parameters. *J. Morph.* 197, 53-62.

Meinhardt, H. (1978). Space-dependent Cell Determination under the control of a morphogen gradient. *J. theor. Biol.* 74, 307-321.

Meinhardt, H. (1982). Models of biological pattern formation. Academic Press, London

Meinhardt, H. (1984). Models for positional signalling, the threefold subdivision of segments and the pigmentation pattern of molluscs. *J. Embryol. exp. Morph.* 83, (Supplement) 289-311.

Meinhardt, H. (1992). Pattern-formation in biology - a comparison of models and experiments. *Reports on Progress in Physics* 55, 797-849.

Meinhardt, H. (1993). A model for pattern-formation of hypostome, tentacles, and foot in hydra: how to form structures close to each other, how to form them at a distance. *Dev. Biol.* 157, 321-333.

Meinhardt, H. and Klinger, M. (1987). A model for pattern formation on the shells of molluscs. *J. theor. Biol.* 126, 63-89.

Mirollo, R.E. and Strogatz, S.H. (1990). Synchronization of pulse-coupled biological oscillators. *SIAM J. On Applied Mathematics* 50, 1645-1662.

Moseley, H. (1838). On the geometrical forms of turbinated and discoid shells. *Philos. Trans. Roy. Soc. Lond.* 351-370

Murray, J.D. (1989). Mathematical biology. Springer, Heidelberg, New York.

Neumann, D. (1959a). Experimentelle Untersuchungen des Farbmusters auf der Schale von *Theodoxus fluviatilis* L. In: Verh. Deutsch. Zoolog. Gesellsch. Münster/Westph. pp.152-156 (Akademische Verlagsgesellschaft Leipzig).

Neumann, D. (1959b). Morphologische und experimentelle Untersuchungen über die Variabilität der Farbmuster auf der Schale von *Theodoxus fluviatilis* L. *Z. Morph. Ökol. Tiere* 48, 349-411.

Neumann, D. (1959c). Musterumschlag auf der Molluskenschale. *Experienta* 15, 178.

Nüsslein-Volhard, C. (1991). Determination of embyonic body axes in the *Drosophila* embryo. *Development* (Supplement) 1, 1-10.

Oppenheimer, P. (1986). Real time design and animation of fractal plants and trees. *Computer Graphics* 20, 55-64.

Pickover, C.A. (1989). A short recipe for seashell synthesis. *IEEE Computer Graphics and Applications*, 9, 8-11.

Pickover, C.A. (1991). Computers and the Imagination. St. Martin's Press.

Plath, P.J., Schwietering, J. (1992). Improbable event in deterministically growing patterns. In: *Fractal Geometry and Computer Graphics* (J.L. Encarnucao, H.O. Peitgen, Skas, G. Englert; Eds.) Springer Verlag, Heidelberg

Prigogine, I. and Lefever, R. (1968). Symmetry breaking instabilities in dissipative systems. *II. J. chem. Phys.* 48, 1695-1700.

Prusinkiewicz, P. (1994). Visual models of morphogenesis. *Artificial Life* 1, 61-74.

Prusinkiewicz, P. and Streibel, D. (1986). Constraint-based modeling of three-dimensional shapes. In: *Proceedings of Graphics Interface '86 - Vision Interface '86*, pp. 158–16.

Ptashne, M., Jeffrey, A., Johnson, A.D., Maurer, R., Meyer, B.J., Pabo, C.O., Roberts, T.M. and Sauer, R.T. (1980). How the lambda repressor and Cro work. *Cell* 19, 1-11.

Raup, D.M. (1962). Computer as aid in describing form in gastropod shells. *Science* 138, 150-152.

Raup, D.M. (1969). Modeling and simulation of morphology by computer. *Proceedings of the North American Paleontology Convention*, pages 71-83

Raup, D.M. and Michelson, A. (1965). Theoretical morphology of the coiled shell. *Science* 147, 1294-1295.

Sabelli, B. (1979). Guide to Shells. Simon and Schuster.

Saunders, B.W. (1984). *Nautilus* growth and longevity: Evidence from marked and recaptured animals. *Science* 224, 990-992.

Segel, L. (1984). Modelling dynamic phenomena in molecular and cellular biology. Cambridge University Press, Cambridge

Segel, L.A. and Jackson, J.L. (1972). Dissipative structure: an explanation and an ecological example. *J. theor. Biol.* 37, 545-549.

Seilacher, A. (1972). Divaricate patterns in pelecypod shells. *Lethaia* 5, 325-343.

Seilacher, A. (1973). Fabricational noise in adaptive morphology. *Systematic Zool.* 22, 451-465.

Stewart, I. (1991). All together now.... *Nature,* 350, 557-557.

Thompson, d'Arcy, W. (1952). On Growth and Form. Cambridge University Press, Cambridge.

Thompson, d'Arcy, W. (1961). On Growth and Form (Abridged Edition). Cambridge University Press, Cambridge

Turing, A. (1952). The chemical basis of morphogenesis. *Phil. Trans. B.* 237, 37-72.

Waddington, C.H. and Cowe, R.J. (1969). Computer simulation of a mulluscan pigmentation pattern. *J. theor. Biol.* 25, 219-225.

Wanscher, J.H. (1971). Considerations on phase-change and decorations in snail shells. *Hereditas* 71, 75-94.

Ward, P.D. and Chamberlain, J. (1983). Radiographic observation of chamber formation in Nautilus pompilius. *Nature* 304, 57-58.

Wilcox, M., Mitchison, G.J. and Smith, R.J. (1973). Pattern formation in the blue-green alga, *Anabaena*. I. Basic mechanisms. *J. Cell Sci.* 12, 707-723.

Willmann, R. (1983). Die Schnecken von Koos. *Spektrum der Wissenschaft (Februar issue) pp.* 64-76.

Winfree, A.T. (1980). The geometry of biological time. Springer Verlag, New York, Heidelberg, Berlin.

Wolfram, S. (1984). Cellular automata as models of complexity. *Nature* 341, 419-424.

Wolpert, L. (1969). Positional information and the spatial pattern of cellular differentiation. *J. theor. Biol.* 25, 1-47.

Index